A Pictorial Guide to Metamorphic Rocks in the Field

A Pictorial Guide to Metamorphic Rocks in the Field

Kurt Hollocher

CRC Press
Taylor & Francis Group
Boca Raton London New York

CRC Press is an imprint of the
Taylor & Francis Group, an **informa** business

CRC Press/Balkema is an imprint of the Taylor & Francis Group, an informa business

Typeset by DiacriTech Limited, Chennai, India
Printed and bound in India by Replika Press Private Limited, Sonipat, Haryana.

Published by: CRC Press/Balkema
P.O. Box 11320, 2301 EH Leiden, The Netherlands
e-mail: Pub.NL@taylorandfrancis.com
www.crcpress.com – www.taylorandfrancis.com

Library of Congress Cataloging-in-Publication Data

Hollocher, Kurt, author.
A pictorial guide to metamorphic rocks in the field / Kurt Hollocher.
pages cm
Includes bibliographical references and index.
ISBN 978-1-138-02630-8 (paperback)—ISBN 978-1-315-76141-1 (ebook) 1. Metamorphic rocks—Pictorial works. I. Title.

QE475.H65 2014
552'.4—dc23

2014024207

ISBN: 978-1-138-02630-8 (Paperback)
ISBN: 978-1-315-76141-1 (eBook PDF)

Dedication

This book is dedicated to all of the world's field mappers, without whom instrument crank-turners in the labs, like me, would have nothing useful to do.

Table of Contents

Preface

This book is a photographic introduction to metamorphic rocks. It is meant to be a survey of mostly typical metamorphic rocks, minerals, and structures as they appear in the field, and the metamorphic processes that make them. Its purpose is to help guide the eye of any student of geology to better see the geologic features in metamorphic rocks that might otherwise remain invisible or obscure. This book is a training tool, one of many, to help budding geologists of metamorphic rocks, young and old.

This book should be accessible to geoscience undergraduate and graduate students, and even to pre-university students if they are interested in rocks, and have spent time reading some introductory geology materials. This book should also be useful for professionals who may be unfamiliar with metamorphic rocks, or those who have started working on them again after a long hiatus. It is likely to be most valuable for people starting work on a new project involving unfamiliar metamorphic rocks, where a visual introduction to rock types, features, and terminology can speed the project along.

Included with the field photos are several line drawings, the purpose of which differs from chapter to chapter. In some cases, they show before and after illustrations of processes that result in particular metamorphic features. In other cases, they represent the features themselves, but more schematically and so, perhaps, more clearly than might appear in a single field photo. In all cases they are meant to complement the outcrop photos.

Also included are images taken from thin sections of metamorphic rocks. Thin sections are 30 μm (micron) thick slices of rock on glass slides, thin enough for most minerals to be transparent. Their purpose is to make the important link between the insights and observational skills that can be learned from looking at metamorphic rocks in thin section, and recognizing and interpreting the same small-scale features in the field. To further this goal, all photomicrographs have field widths of 4 mm. At this magnification, many features visible in thin section are translatable to features in the field that can, with training and experience, be seen with a good rock surface and a 10x hand lens.

Like all aspects of geology, metamorphic rocks pose an intellectual challenge. The immediate goal of working in metamorphic terranes is usually to figure out the rock types, map patterns, and structures, and to work out the local geologic history. This understanding can be used for academic purposes, such as planning a petrologic or geochemical sampling program. Ultimately the goal may be reconstructing how entire ancient mountain ranges developed (may require additional information). Alternatively, the preliminary work may be applied to searches for new ore deposits, or underground extensions of known deposits. Understanding metamorphic rocks and structures may help in the design and construction of tunnels through such terrane. Conversely, excavations can give access to rock exposures that are otherwise inaccessible. The point is that metamorphic rocks are widespread and important beyond academic geology. They are also fun to work in.

About the author

I was first exposed to rocks and geology as a lad in the early 1960's, on family vacations to the coast of New England and the Canadian Maritime provinces. My father would tell me things about the rocks, which ran the gamut from fossil-rich sedimentary to igneous and metamorphic rocks. One of my oldest memories was of asking my father if the deepest parts of the oceans was in their centers, reasoning that those areas were farthest from the continental sediment sources (yes, I was a strange one). He answered that, no, the deepest places were narrow trenches, generally quite close to land. That answer made me realize that the Earth must be a more interesting place than I had realized, as developments in plate tectonic theory soon demonstrated.

As I grew up, we traveled to northern Europe and hiked extensively in the New England Appalachians. I took many photographs using a series of progressively less-primitive cameras, commonly of geologic features that seemed interesting. In college and graduate school I learned better what geologic features actually were interesting, and continued photographing. In graduate school I worked on metamorphic rocks in southern New England. Then and since I attended numerous field trips, both ad hoc, during research work, and associated with different geologic organizations. After graduate school I was hired at Union College, a small liberal arts undergraduate school in east-central New York State. In that environment I taught fledgling earth scientists, and others, about the solid Earth and its internal and external processes. That forced me to wade through decades of old photos, and to take many new ones, to illustrate how our world is constructed and how it works. A subset of those photos makes up most of the material in this book.

Many of my friends will probably wonder what possessed me to write this book. I have missed altogether too many field trips. I am not a world-class petrologist, and I am certainly no structural geologist. I am not a good field mapper, and in fact I was once told by my Ph.D. advisor to quit wasting my time trying to do mapping. It turns out that I am a pathological splitter of geologic units, rather than a lumper. My mapped units never spanned more than an outcrop or two, and so never made any large-scale sense. I am also not a professional photographer and have never had access to equipment of more than moderate professional quality. A large fraction of my photos were taken in typical temperate zone conditions, which usually included some combination of dark woods, twilight, and rain. Fortunately, quantity has a quality all its own. From the large assortment of slides, negatives, and electronic images, there were some pictures that were actually pretty good, or at least showed an interesting feature or two.

The thought of writing a book, even a picture book, had never crossed my mind. One fateful day, however, I was contacted by Robert Nesbitt, a geologist at the University of Southampton. Apparently he noticed that I had a lot of geology photos on my web page, and he was looking for someone from the New World to help write a picture book on igneous and metamorphic rocks in the field. Over years of delays caused by more pressing work, and occasional advances, we eventually decided to separate the topics into two books. I got the metamorphic part. My motivation for this book was simply the idea that someone else thought I could contribute to such a thing. And here it is.

Acknowledgments

I would like to thank my many teachers and colleagues who, over the years, have given me the training, insight, and opportunities to see and work with metamorphic rocks in the field. This includes field trip leaders for many excursions to metamorphic terranes, many of which were under the auspices of organizations like the New England Intercollegiate Geological Conference, the New York State Geological Association, the Geological Society of America, the International Geological Congress, and the International Mineralogical Association. Individuals are too numerous to mention in full, but include, in alphabetical order: Henry N. Berry IV, John B. Brady, Jack T. Cheney, Robert H. Fakundiny, John I. Garver, David Gee, Charles V. Guidotti, Leo M. Hall, Lindley S. Hanson, Norman L. Hatch, J. Chris Hepburn, Rudolph Hon, Arthur M. Hussey II, Jonathan Kim, Jo Laird, James M. McLelland, Phillip H. Osberg, William Peck, Stephen G. Pollock, Douglas W. Rankin, Nick M. Ratcliffe, David Roberts, Frank S. Spear, Barbara Tewksbury, James B. Thompson, Peter J. Thompson, Phillip R. Whitney, Robert P. Wintsch, and, most especially, Peter Robinson (the one from the University of Massachusetts, Amherst, and Norwegian Geological Survey, Trondheim). To these people, and many others, I am thankful. I thank Robert W. Nesbitt, for having started me on this book project, and for letting me use some of his photos (Figs. 13.9, 15.5, 15.6, and 18.4). I also thank John C. Schumacher, Thomas C. Hollocher, and my wife and daughter, Janet and Alice, for their helpful reviews of this manuscript.

A matter of scale

With rocks it is typically difficult to tell how big things are without some context. An image may show metamorphic garnets, but are they 1 mm across, or 100? Is the fold pictured a few centimeters from limb to limb, a few meters, or does it cover a major part of a mountain face? In some studies of metamorphic rocks the exact scale matters, for example a study of mineral size variation with metamorphic grade. In such circumstances you need to measure mineral sizes in many different places, and photos should probably include an accurate length scale, such as a ruler. In this book accurate length scales are not necessary, but it is still helpful for the viewer to have some idea of how big things are. Otherwise, it might be difficult to judge how big to expect similar

Figure i Examples of scale indicators used in the field photography of this book.

things to be in the field. Features might be missed simply by the accident of looking at the wrong scale. Almost all of the field pictures shown in this book have some sort of scale indicator, some examples of which are shown in Figure i. Although a ruler-type scale might be ideal, given forgetfulness and the press of time on the outcrop, they are not always quickly available, or available at all. Often, the quickest thing handy is used as a size indicator: a rock hammer, a pencil or field notebook, or a coin. When time is very short, fingers, rubber boots, plants, or even dead leaves may be pressed into service. Generally the scale items seen in this book are widely familiar, so readers who have had their eyes open for awhile should easily be able to estimate the scales to within a factor of two or so. That is good enough for most work, and fine for the purposes of this book. In the few cases where visual scales have been neglected entirely, image field widths are given in the captions.

Chapter 1

Introduction

THE BASICS

If the Earth's water were removed, most of its surface would be covered by sediments and sedimentary rock. Like icing on a cake, however, this thin veneer merely covers a voluminous and equally interesting interior: hot deep crust, a slowly convecting mantle, rising magmas, and sinking oceanic plates. Using round numbers, the silicate Earth makes up 84% of the planet's volume. Because the sedimentary cover is so thin, and only a minute fraction of the upper mantle and crust are molten at any given moment, and because deformation and other solid state processes have modified most older sedimentary and igneous rocks, most of our planet is made of metamorphic rock.

The classic way to think about the origin of metamorphic rock is to start with the protolith concept. A protolith, from the Greek meaning first stone, is the precursor rock that existed before any metamorphic change. Protoliths can be any of the many varieties of igneous or sedimentary rocks, such as basalt, granite, shale, or sandstone. Metamorphic conditions cause the rocks to change their original form in terms of shape, mineralogy, grain size, and even chemical composition. The boundaries separating the realm of metamorphism from the realms sedimentary and igneous rocks are somewhat arbitrary, governed more by historical precedent than any obvious or objective criteria.

The word metamorphism comes from Greek, meaning to change form. Rocks can change form in many ways. On the cold side, loose sediments gradually transform to sedimentary rocks. Burial results in higher temperatures and pressures. These new conditions, along with other processes, cause pore spaces to be filled in, allow replacement of feldspars by clay, and drive reactions that modify clay mineralogy. These transformations, while metamorphic in a sense, result in materials that still look like sediments, and so are generally worked on by people who specialize in sedimentary rock geology. These transformative processes are usually referred to as sediment diagenesis, rather than metamorphism.

On the hot side, high temperatures can result in melting, where the rocks transition, generally only in part, to silicate liquid (magma). A partially melted rock has a solid metamorphic part (restite) and an igneous liquid part that occurs initially along grain boundaries. The liquid can separate from restite by migrating into nearby opening spaces, or, because it is generally less dense than the solid restite, the liquid can migrate upwards along grain boundaries. Separation of magma from the solid residue may form dikelets, which can coalesce into dikes as the liquid moves. Magmas may pond and accumulate at some storage or emplacement level to form magma chambers, and some may escape to the surface to feed volcanoes.

Magmas eventually cool, crystallize, and become solid igneous rocks. As they crystallize they may undergo other changes as well. These changes can include recrystallization, unmixing of solid solutions, and mineral alteration caused by reactions between igneous minerals and late fluids that are commonly left over from magma crystallization. Fluid-mineral reactions can form alteration products, such as sericite (fine-grained white micas) in feldspars, and chlorite, which may partly replace biotite. These changes, while metamorphic in a sense, are traditionally studied by people who specialize in igneous rocks and so are considered to be part of the igneous realm.

Metamorphic rocks therefore form between fuzzy lines that separate them from the realms of sedimentary and igneous rocks. Such a seemingly restricted domain, however, turns out to be quite diverse and complex (remember the 84%). Because most of the Earth is inaccessible, this book will concentrate on metamorphic rocks found in the crust. Through tectonic happenstance, some mantle rocks are lucky enough to become incorporated into the crust and reach the surface, where we can see them. The conditions and processes that produce metamorphic rocks can be divided into temperature, pressure, strain (deformation), chemical flux, and recrystallization, which will briefly be discussed in sequence.

PRESSURE, TEMPERATURE, AND METAMORPHIC GRADE

Pressure and temperature (P and T) are discussed together because they commonly change together. With depth in the Earth's interior, pressure increases as a result of the weight of overlying rock (a product of rock density, gravitational acceleration, and depth). If we ignore highly porous rocks, common rocks range in density from about 2500 kg/m^3 for shales, and 2900 kg/m^3 for typical crust, to 3400 kg/m^3 for typical upper mantle rock. These densities correspond to depths of 4000, 3500, and 3000 m, respectively, to produce 1000 bars (1 kbar) of lithostatic pressure. As a rule of thumb, pressure therefore increases by 1000 bars for each 3.5 km of depth in the crust (multiply bars by 100,000 to get Pascals, or divide by ten to get megaPascals).

As another rule of thumb, the near-surface geothermal gradient is about 20°C/km in typical continental interiors. Multiplying the two rules of thumb, one gets temperature gradients of about 70°C/kbar. Geothermal gradients vary considerably, but if one avoids regions directly over magma chambers and the coldest subduction zones, the range is something like 10-40°C/km or 35-140°/kbar. Although near-surface temperature gradients tend to look straight, with depth they become concave downward (concave toward higher pressure), as temperatures increase progressively more slowly with depth. Another rule of thumb is that one should be wary of extrapolating too far.

Rocks that are out of equilibrium with their environment undergo changes to reestablish equilibrium. Take a piece of shale, for example. This rock has a particular chemical composition. At low temperatures and pressures the chemical components might occur as clay and quartz, the (more or less) stable mineral assemblage. At higher temperatures and pressures the chemical components tend to become rearranged as other sets of minerals become stable. Minerals transform from one set to another via a series of chemical reactions. These reactions can cause some minerals to vanish and others to appear in the rock as conditions change. Minerals can also remain present, but can change in abundance or composition. For any particular rock (chemical composition) there will be a stable set of minerals (mineral assemblage) at every metamorphic temperature and pressure. Minerals react to approach equilibrium as the conditions change. We will not worry about the fact that minerals in real rocks were rarely ever in perfect equilibrium with one another.

Some common rocks, like typical basalts and shales, have limited ranges of chemical composition. This means, for example, that a basalt metamorphosed at one set of metamorphic conditions will tend to have the same mineralogy as a similar basalt metamorphosed at the same conditions anywhere in the world. The consistent chemical composition results in a consistent pattern of metamorphic minerals over the range of P-T conditions (an oversimplification, but close enough for now). For this reason, typical basalts yield reasonably consistent mineralogy that can be a general guide to the metamorphic conditions that the rock experienced. Using the mineral assemblages in metamorphosed basalt at different P-T conditions, we can divide P-T space into a series of fields where basalts generally have the same mineral assemblages (Fig. 1.1). These fields are named in different ways, but they generally involve the rock type, what the rock

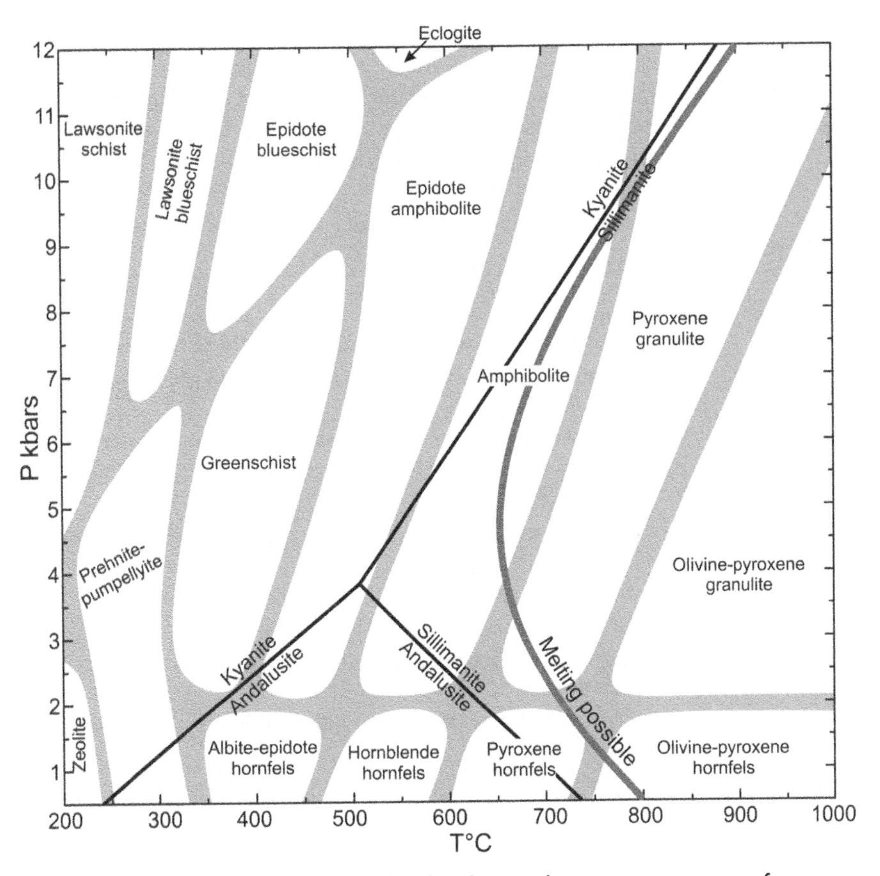

Figure 1.1 Metamorphic facies schematic for basaltic rocks across a range of pressures and temperatures found of the crust. Note that the aluminosilicate reaction lines are for reference purposes in this diagram, because basaltic rocks are not generally expected to have aluminosilicates in this P-T range. Diagrams such as this one are useful guides for the likely metamorphic conditions experienced by common rock compositions (Table 1.1), but should never be thought of as applicable to all rocks. Even here, basaltic rocks that have chemical compositions moderately different from the average can have somewhat different mineral assemblages at the same metamorphic conditions. This diagram is based on information from several sources, including Perple-X thermodynamic modeling (Connolly, 2009) of several average basalt compositions saturated with pure H_2O fluid, the low-temperature phase relations of Evans (1990), Baltatzis (1996), Muñoz et al. (2010), and Tsai et al. (2013), and the low-pressure fields of Winter (2001, fig. 25-2) and Blatt et al. (2006, fig. 17-4).

looks like, and/or the characteristic minerals they contain. The fields are typically referred to as metamorphic facies (the most common term used for metamorphosed basaltic rocks), or grades or zones (more commonly used for metamorphosed shales and some other rocks). They all mean the same thing, however.

For example, basaltic rocks in the greenschist field (Fig. 1.1) typically contain chlorite, white mica, and a few other minerals that are stable at those conditions. Chlorite gives the rocks a green color, and the platy minerals chlorite and white mica can be oriented parallel to one another to give the rock a schist-like appearance. Hence, greenschist facies. Basaltic rocks in the pyroxene granulite field typically contain two pyroxenes (augite and orthopyroxene) and plagioclase, but are mica-poor. The scarcity of platy minerals like the micas, and generally blocky shapes of pyroxenes and plagioclase, cause many of these rocks to be textural granulites. Hence, pyroxene granulite facies. The change in mineralogy and rock appearance from greenschist to pyroxene granulite facies occurs because minerals like chlorite and the micas react to form a different stable assemblage of feldspar and pyroxenes. Figure 1.1 can therefore be used as a general guide to the metamorphic conditions experienced by typical basaltic rocks. Table 1.1 lists the minerals that are commonly found in basaltic rocks that have been

Table 1.1 Typical minerals in metamorphosed normal basaltic rocks in the metamorphic facies fields in Figure 1.1.

Facies	*Typical minerals**
Zeolite	Analcite and other zeolites, prehnite, zoisite, white micas, albite, calcite
Prehnite – pumpellyite	Prehnite, pumpellyite, chlorite, albite, actinolite, epidote, white micas, calcite
Lawsonite	Lawsonite, chlorite, actinolite, phengite, aragonite, garnet
Lawsonite blueschist	Glaucophane, lawsonite, garnet, albite, chlorite, zoisite, phengite, aragonite
Epidote blueschist	Glaucophane, garnet, epidote, albite, chlorite, zoisite, omphacite, phengite, aragonite
Eclogite	Garnet, omphacite, pargasite, zoisite, kyanite, phengite
Greenschist	Chlorite, albite, white micas, actinolite, calcite
Epidote amphibolite	Plagioclase, hornblende, epidote, actinolite, biotite, chlorite, calcite, garnet
Amphibolite	Plagioclase, hornblende, cummingtonite, biotite, garnet, clinopyroxene, orthopyroxene
Pyroxene granulite	Plagioclase, clinopyroxene, orthopyroxene, garnet, biotite
Olivine – pyroxene granulite	Plagioclase, clinopyroxene, orthopyroxene, olivine, garnet
Albite – epidote hornfels	Albite, epidote, actinolite, chlorite, white micas
Hornblende hornfels	Hornblende, plagioclase, orthopyroxene, clinopyroxene, epidote, garnet
Pyroxene hornfels	Clinopyroxene, orthopyroxene, plagioclase, garnet
Olivine – pyroxene hornfels (Sanidinite**)	Clinopyroxene, orthopyroxene, plagioclase, garnet, olivine

* Quartz, or other silica polymorph, can occur in any facies, depending on rock composition, though not with Mg-rich olivine.

** Rocks rich enough in potassium and silica may have sanidine and tridymite, though preservation of them requires rapid cooling as well as high metamorphic temperatures.

metamorphosed in the various facies fields. The broad boundaries between the fields are caused by a number of factors, but especially compositional variations in the basaltic protoliths themselves.

Shales are another good example of a common rock that has, globally, a reasonably consistent chemical composition. Because of this, the mineralogy of metamorphosed shales also reflects, in general, the metamorphic conditions the rocks experienced (Fig. 1.2). What is somewhat remarkable about metamorphosed shales is that they can form a set of easily recognizable minerals over a relatively narrow temperature range

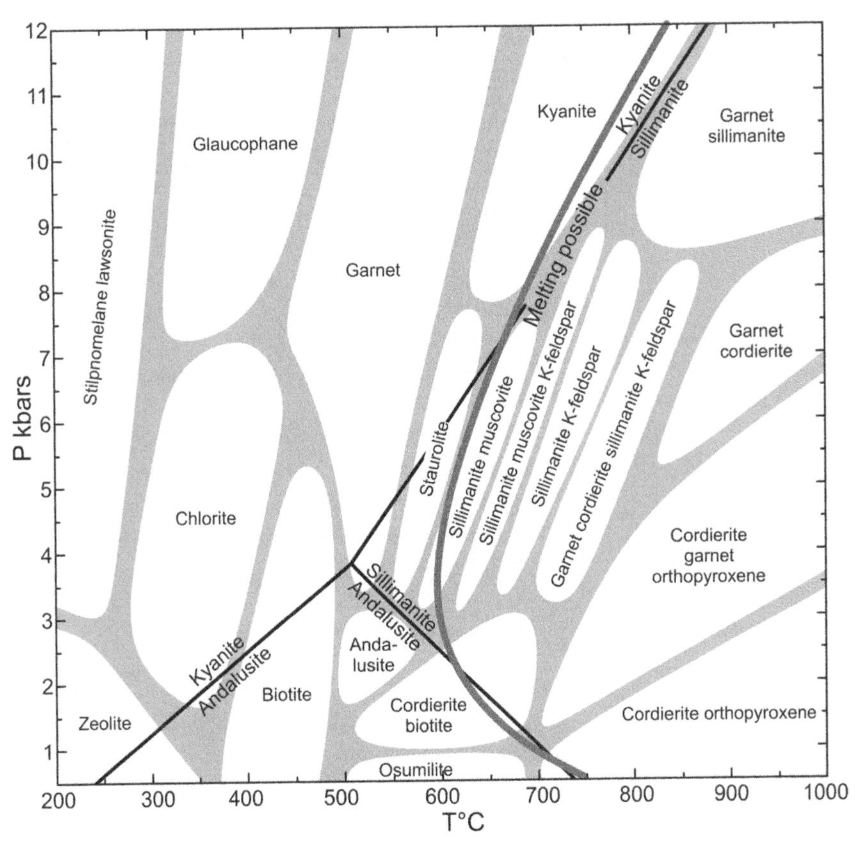

Figure 1.2 Metamorphic grades for pelitic rocks across a range of temperatures and pressures found of the crust. This diagram can be used as a guide to the metamorphic conditions experienced by shales, provided you pay attention to all minerals in the assemblage (Table 1.2). Unfortunately, as with basalts, the chemical compositions of shales are not all the same, and so they can have somewhat different assemblages than those shown here. For example, in the garnet – cordierite – sillimanite – K-feldspar zone, a rock with enough additional calcium or sodium may have more feldspar and no sillimanite. One with more magnesium may have no garnet, and one with more iron may have no cordierite. On the other hand, mineralogical differences are one of the keys to mapping geologic units. This diagram is based on information from several sources, including Tracy et al. (1976), Westphal et al. (2003), and El-Shazly et al. (2011), and calculations with the Perple-X thermodynamic modeling software (Connolly, 2009), using a modified marine shale composition (Na_2O, 0.96% by weight; MgO, 3.19; Al_2O_3, 18.13; SiO_2, 65.00; K_2O, 4.39; CaO, 0.48; MnO, 0.13; FeO, 7.72), and assuming saturation with pure H_2O fluid.

in the middle of the diagram, where many regionally metamorphosed crustal rocks form. Table 1.2 is a list of the minerals that can be found in metamorphosed shales at different grades.

With a little recollection of mineralogy, you should be able to see in Figures 1.1 and 1.2 and Tables 1.1 and 1.2 that, in general, low temperature assemblages are rich in hydrous minerals like chlorite and micas, whereas high temperature rocks are dominated by anhydrous assemblages rich in garnet, pyroxenes, cordierite, and feldspars. Higher pressures tend to stabilize hydrous minerals to higher temperatures.

Table 1.2 Typical minerals found in metamorphosed shales, in addition to quartz, in the metamorphic grade fields in Figure 1.2.

Metamorphic grade	*Typical minerals**
Stilpnomelane – lawsonite	Stilpnomelane, lawsonite, white mica, albite, chlorite, kaolinite, garnet, calcite or aragonite
Glaucophane	Glaucophane, phengite, garnet, albite, chlorite, aragonite, lawsonite
Garnet	Garnet, biotite, albite, chlorite, white mica
Kyanite	Garnet, kyanite, staurolite, albite, biotite, muscovite or phengite
Garnet – sillimanite	Garnet, sillimanite, K-Na-feldspar
Zeolite	Zeolites, albite, chlorite, white micas, calcite, stilpnomelane
Chlorite	Chlorite, white mica, epidote, albite
Biotite	Biotite, chlorite, albite, white mica, epidote
Staurolite	Staurolite, garnet, biotite, plagioclase, muscovite
Sillimanite – muscovite	Sillimanite, garnet, biotite, plagioclase, muscovite
Sillimanite – muscovite – K-feldspar	Sillimanite, garnet, biotite, plagioclase, muscovite, K-feldspar
Sillimanite – K-feldspar	Sillimanite, garnet, biotite, plagioclase, K-feldspar
Garnet – cordierite – sillimanite – K-feldspar	Sillimanite, garnet, cordierite, biotite, K-feldspar
Garnet – cordierite	Garnet, cordierite, K-feldspar, sillimanite, plagioclase orthopyroxene, biotite
Andalusite	Andalusite, staurolite, garnet, biotite, plagioclase, muscovite
Cordierite – biotite	Cordierite, biotite, muscovite, plagioclase, K-feldspar, garnet
Osumilite	Osumilite, cordierite, plagioclase, K-feldspar, orthopyroxene
Cordierite – garnet – orthopyroxene	Orthopyroxene, cordierite, plagioclase, K-feldspar, garnet, biotite
Cordierite – orthopyroxene (Sanidinite**)	Cordierite, orthopyroxene, plagioclase, K-feldspar

* Quartz is stable in all fields, white micas refers to muscovite with or without paragonite (low-grade only), or phengite at high-pressure and low-temperature.

** Rocks rich enough in potassium and silica may have sanidine and tridymite, though preservation of them requires rapid cooling as well as high metamorphic temperatures.

MOVEMENT OF ROCKS THROUGH P-T SPACE

Rocks, of course, do not simply appear at a particular metamorphic P-T condition and then rush, nicely equilibrated, to your outcrop. Metamorphic rocks have to travel slowly through a range of P-T conditions by typically starting out at or near the surface of the Earth, being buried and heated, and then cooling and decompressing while returning to the surface where we can see and collect them. This is referred to as a P-T path. Figure 1.3 shows a simple P-T path. The rock starts at the surface, perhaps as a lava flow. Gradually it is buried by other lava flows and sediments, overriding thrust sheets, and horizontal crustal shortening (during which the thickening crust pushes all parts to greater depth). As burial progresses, the rocks heat up as a result of heat flow from hotter rocks below, from radioactive decay in the rocks themselves, and from the insulating effect of the overlying pile of rock. Rising magmas and some deformation processes can add additional heat. Though there are heat sources, remember that many metamorphic reactions, especially dehydration reactions and melting, are endothermic and so can limit temperature rise to some extent. The part of the P-T path in the direction toward higher temperatures is typically referred to as the prograde metamorphic path.

The processes that raise mountains have limits. The higher the mountains rise, the faster erosion, lateral tectonic escape, or detachment faulting tend to limit mountain height. Orogenic forces reach equilibrium with unroofing processes at some point, which may be at approximately the same time as the maximum pressure experienced by a rock in the mountain belt. As the orogenic forces wane, from cessation of subduction and slab detachment, for example, rock heating can continue even as unroofing processes start to bring rocks to the surface. This allows them to heat and decompress

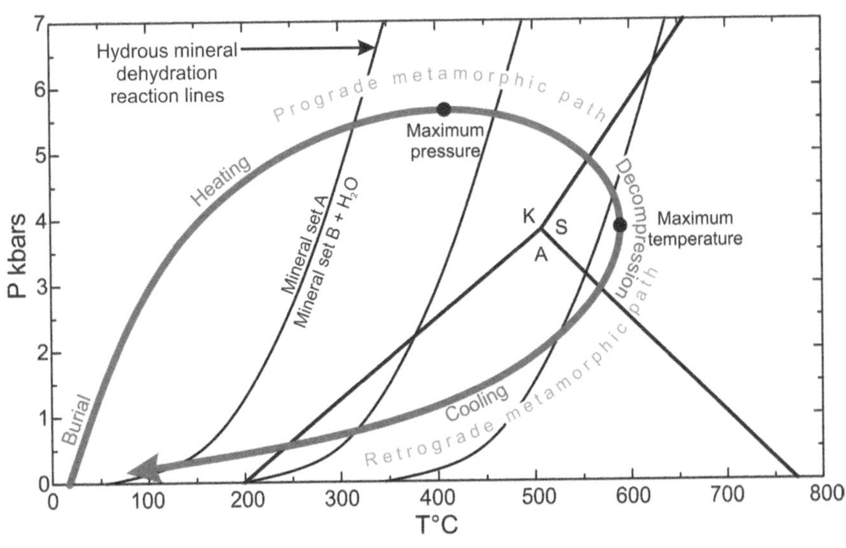

Figure 1.3 Simplified P-T path that might be taken by a rock during regional metamorphism, from the surface, through different metamorphic pressures and temperatures, and back to the surface again. Dark blue, curved lines represent generic dehydration reactions. The solid black lines are aluminosilicate reaction lines, with the aluminosilicate stability fields labeled K, S, and A for kyanite, sillimanite, and andalusite, respectively.

simultaneously for awhile. Eventually the rocks will pass through a point of maximum temperature, and finally unroofing brings the rock to the cold surface where we break it off the outcrop. The part of the P-T path from the point of maximum temperature to the surface is usually referred to as the retrograde metamorphic path.

PRESERVING THE PROGRADE ASSEMBLAGE

Figure 1.3 raises an important question. If the metamorphic assemblage in a rock represents a close approach to equilibrium, why do we see metamorphic assemblages at the surface at all? If a shale starts out as a mixture of clay and quartz, and the minerals progressively reequilibrate as the rock follows the P-T path, why don't we end up with a reequilibrated mess of clay and quartz at the end? The answer comes in two parts.

First, reactions between minerals take time, generally a lot of it (reaction kinetics), especially if the rocks are not actively deforming. During prograde metamorphism, temperatures are initially low, and reactions tend to be slow. Mineral assemblages may not reach equilibrium, though fine grain size and deformation-induced recrystallization may help. Because reaction rates increase enormously with increasing temperature, where would you expect the metamorphic reactions to be the fastest and the assemblage to be closest to equilibrium? Near the point of maximum temperature, of course. Because of fast reaction rates, including solid-state diffusion within minerals, equilibration near the point of maximum temperature tends to partially or wholly erase evidence of earlier metamorphic history, especially at medium- and high-temperatures.

The second part comes from the metamorphic reactions themselves. As a rock follows the prograde path through the crust, it crosses numerous reaction lines that include dehydration reactions. A shale may cross, say, chlorite-out, paragonite-out, and staurolite-out reaction lines, or others that only partially remove these hydrous minerals. As these minerals decompose, what happens to the released H_2O? Some of it may go into other minerals of the new assemblage, such as biotite, but much of it enters the grain boundary fluid phase that is in equilibrium (more or less) with the surrounding minerals. Fluids have low densities compared to the adjacent minerals, so an extensive network of grain-boundary fluids will tend to hydraulically fracture its way upwards, escaping the rock and thus dehydrating it. As the rock travels along the P-T path, eventually it passes beyond the point of maximum temperature and has lost all of the H_2O from hydrous minerals that it can. Prograde reactions cease, and the rock begins the retrograde segment of its path.

As you can see in Figure 1.3, the retrograde path crosses backwards across the same (or similar) reaction lines that it passed forwards across during prograde metamorphism. Going backwards across a prograde dehydration reaction, while following the retrograde path, makes it a retrograde hydration reaction. What do these reactions need to grow the new, lower temperature, hydrous mineral assemblages? H_2O from a fluid phase, which is no longer available because it has escaped. Indeed, any traces of grain boundary fluid, remaining after the point of maximum temperature, would quickly be used up by the first retrograde hydration reaction encountered, leaving an effectively dry rock. No fluid means no hydration reactions, and also no grain boundary fluid that would speed diffusion transport of the chemical components needed for the reactions. Lack of fluid also inhibits rock recrystallization. Rapid equilibration at maximum temperature, loss of fluid released by dehydration reactions, and reduction in or cessation of deformation, therefore tend to preserve the mineralogy produced near the point of maximum metamorphic temperature.

METAMORPHIC FIELD GRADIENT

If, in general, a metamorphic rock best records the mineral assemblage formed at maximum temperature, what we see on the ground in terms of metamorphic grade is what is called the metamorphic field gradient. That is the maximum temperature, or metamorphic facies, grade, or zone, recorded by different rocks across a metamorphosed region. Figure 1.4 illustrates a regional metamorphic field gradient with a series of simple, nested P-T paths. The points of maximum temperature on the paths are encircled to show the metamorphic field gradient: the range of common P-T conditions recorded by metamorphic assemblages that one typically finds in the field.

One of the many goals of geologists working on metamorphic rocks is to understand how mountain belts work. To do that, you need more than just maximum temperature P-T conditions and the metamorphic field gradient. What you would like is the whole P-T path over time for each rock. Where does that information come from? Well, some of it comes from the metamorphic field gradient, because that is an important series of points on the P-T paths. Some of it can come from structural geology, which can show how rocks were deformed. Some of it can come from radiometric dating, which can show when some minerals grew, when their radiometric clocks were reset, or when they stopped

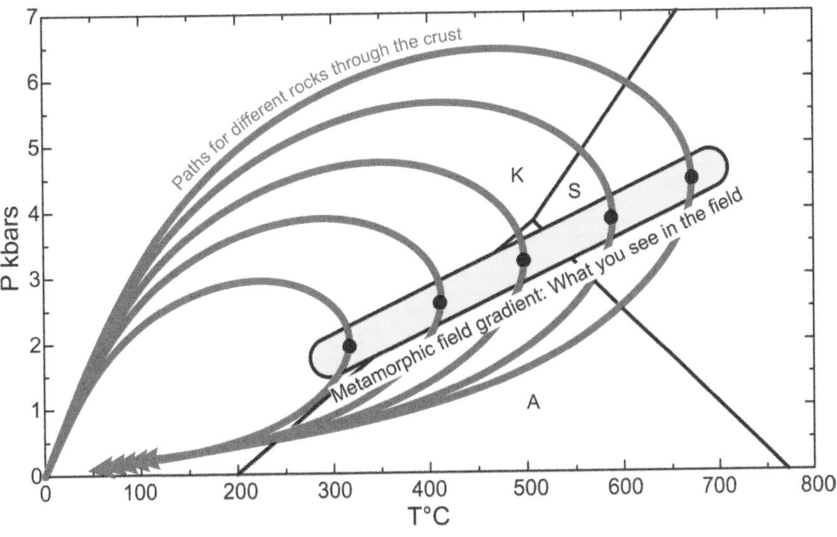

Figure 1.4 A set of nested, simplified P-T paths taken by different rocks in different parts of a regional metamorphic belt. Metamorphic rocks tend to best preserve the mineral assemblage that was stable near the point of maximum temperature (black dot for each path). Collectively these define the metamorphic field gradient, which is the set P-T conditions recorded by the dominant, preserved assemblages found in rocks out in the field. Note that the metamorphic field gradient in no way mimics the P-T paths. K, S, and A are kyanite, sillimanite, and andalusite stability fields, respectively.

being reset as the rocks cooled. Some also comes from thermodynamic modeling and published phase diagrams. The more closely these methods can be coupled to detailed petrographic interpretation of the rocks, done using optical and electron microscopy and microbeam chemical analyses, the richer the yield of P-T path information.

Figure 1.5 is a simple example of how metamorphic mineralogy and textures, determined by optical microscopy, can be useful for determining other parts of a P-T path. The maximum temperature prograde assemblage in Figure 1.5 is preserved outside of the pink garnet, and includes garnet, quartz, plagioclase, biotite, muscovite, kyanite, staurolite, rutile, and ilmenite. Inside the garnet is an assemblage that lacks staurolite and kyanite but has chloritoid. The inclusions are parts of an earlier, lower temperature, possibly higher pressure metamorphic assemblage that was trapped by the growing garnet, and preserved. Figure 1.6 illustrates the kind of information that can be extracted from metamorphic rocks to understand more of the P-T path, and the timing of when the rocks experienced parts of that path. It is unlikely that all of the information shown in Figure 1.6 can be extracted from a single rock, but a collection of different rocks, and a team of people and equipment needed to measure and interpret them, can help produce the paths later used to construct an orogenic model.

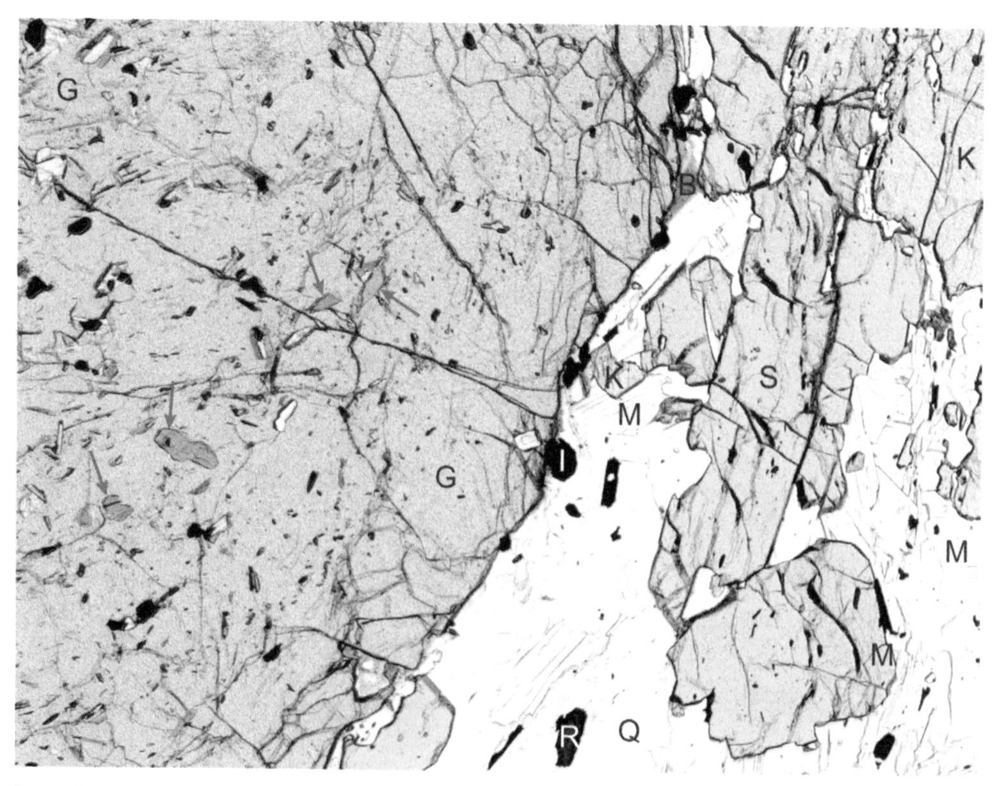

Figure 1.5 This is an image of a pelitic schist in thin section, showing chloritoid inclusions (red arrows) inside garnet that contrasts with the chloritoid-free, staurolite – kyanite assemblage outside of the garnet. The inclusions represent preserved parts of an assemblage that was present early in the rock's metamorphic history, and so provides information on the P-T conditions at the time the chloritoid crystals were overgrown by garnet. The field width is 4 mm, and the image was taken in plane-polarized light. Abbreviations: B, biotite; G, garnet; I, ilmenite; M, muscovite; Q, quartz; R, rutile; S, staurolite. Gassetts, Vermont, USA.

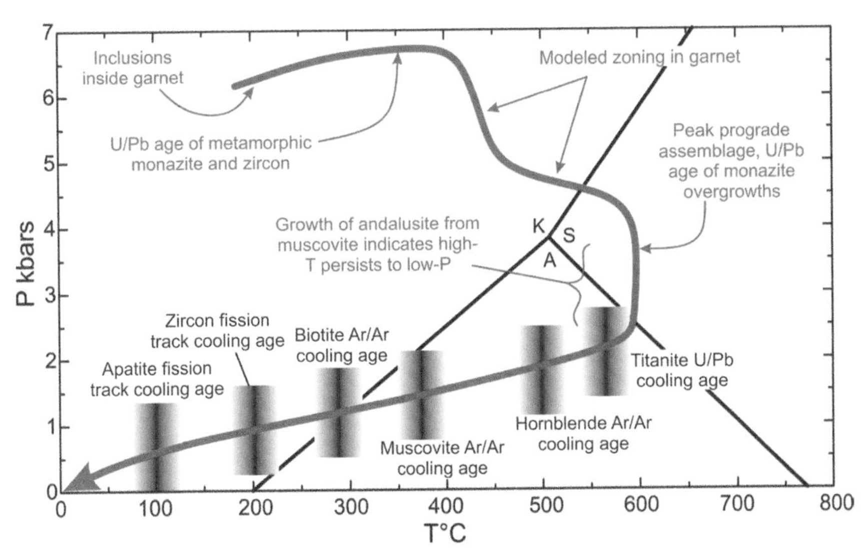

Figure 1.6 Hypothetical example of different ways to extract pressure, temperature, and time information for different parts of a metamorphic rock P-T path. In general, extracting this kind of information can't all be done on a single rock. This sort of work generally requires a variety of rocks, a range of analytical techniques and instruments, and a team to help with the work. For example, pelitic rocks are good sources of monazite, muscovite, and best for interpreting garnet zoning. In contrast, mafic rocks are better sources of titanite, hornblende, and zircon that grew during metamorphism. K, S, and A are kyanite, sillimanite, and andalusite stability fields, respectively.

ROCK STRAIN: FOLIATION AND LINEATION

After temperature and pressure, rock deformation is the most obvious process that has affected most metamorphic rocks. Deformation results in changes to rock shape, the result of which can be seen in deformed fossils or crystals, reorientation of crystals to produce mineral lineations and foliations, and folding. Figure 1.7 shows what happens to rocks that undergo isotropic compression (same in all directions) and differential compression. If the compressional forces are all equal (Fig. 1.7A), there is no change in rock shape, though the volume decreases a little (density increases) and higher pressures may stabilize different minerals (generally more dense ones). If compression in two directions is the same and the third is less (Fig. 1.7B), the rock will extend, or extrude, along one axis, developing a lineation as elongate minerals re-orient themselves parallel to the extension direction. If compression in two directions is less than in the other (Fig. 1.7C), flattening results, along with development of a foliation as platy minerals become aligned parallel to the two extension directions, but no lineation develops. If compressional forces in all three directions are different (Fig. 1.7D), there is differential extension and contraction that results in both a foliation and a lineation (see Chapter 15 for more discussion of this topic).

Figures 1.7E-H show side views of any of Figures 1.7B, C, or D, where round markers are shortened parallel to the maximum compression direction, and extended parallel to the maximum extension direction. Platy or elongate grains also rotate into parallelism with the extension direction. Figure 1.8 shows the alternative way to deform rocks, by simple shear, which operates like a sliding deck of cards. During simple shear, foliation and lineation develops the same way as by pure shear, by grain rotation and by elongation of markers into parallelism.

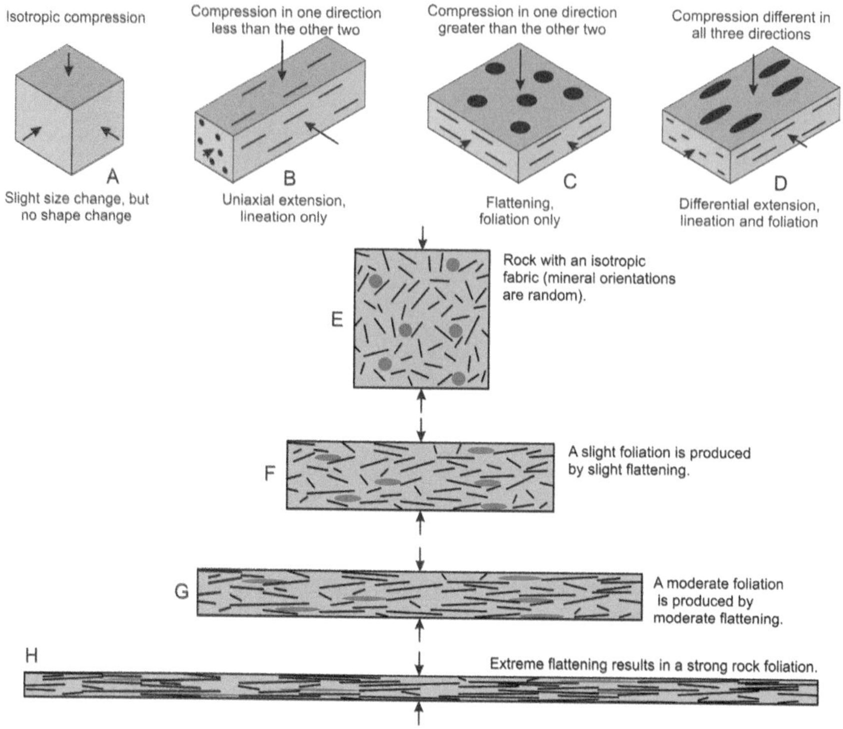

Figure 1.7 Illustration of how differential compression (flattening, pure shear) can change rock shape and re-orient minerals. Changes in shape result in the development of lineation and foliation by deforming marker features and rotating elongate or platy grains. A-D) Shape changes and foliation and lineation differences that result from different amounts of triaxial compression, which are schematically proportional to the arrow lengths. E-H) Side views of a single block (B, C, or D) that is deformed by flattening, with progressive foliation development.

CHEMICAL FLUX

During prograde metamorphism, some reactions involve dehydration of a hydrous assemblage to produce a less hydrous assemblage. An example of such a reaction is:

$$\underset{\text{muscovite}}{KAl_2Si_3AlO_{10}(OH)_2} + \underset{\text{quartz}}{SiO_2} = \underset{\text{sillimanite}}{Al_2SiO_5} + \underset{\text{K-feldspar}}{KAlSi_3O_8} + \underset{\text{fluid.}}{H_2O} \quad (1.1)$$

H_2O can form a low-density fluid phase which, in extensive grain-boundary networks, tends to migrate upward and escape. This means that the rocks undergoing dehydration change chemical composition by losing H_2O. The rocks above may also change composition by gaining H_2O, if hydration reactions produce more hydrous minerals than were present before. More than that, at high temperatures and pressures fluids can dissolve substantial quantities of material, which is carried along with the escaping fluid. As the fluid migrates to regions with lower pressures and temperatures, the dissolved solids can precipitate, thus changing the composition of both source and destination rocks for components other than just H_2O (Fig. 1.9A). Figure 1.9B shows a similar situation in which hot fluids escaping from an igneous intrusion carry chemical components

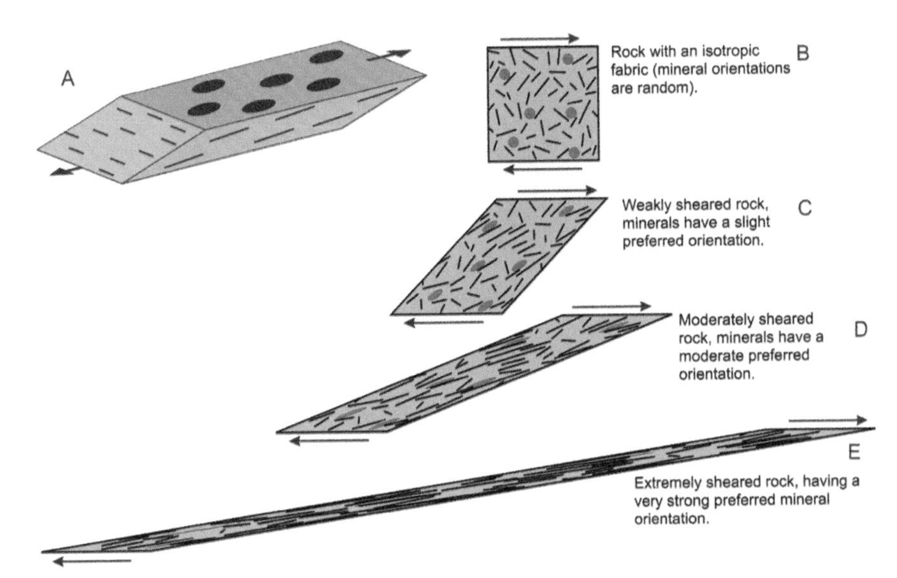

Figure 1.8 Illustration of how simple shear (like the sliding of a deck of cards) can change rock shape and generate foliation and lineation by deforming marker features and rotating platy or elongate grains. A) Shear across a 3-d block. B-E) Side view of deformation of a single block, with progressive foliation development.

(Si, Al, Fe, Mg, Na, H_2O) into adjacent calcite marble, promoting decarbonation of calcite and growth of amphiboles at and near the pluton contact.

Diffusion is another mechanism of chemical flux. Figure 1.9C shows an amphibolite in contact with marble. During metamorphism, with or without grain boundary fluid, chemical components can diffuse in both directions across the contact, with each component diffusing down its own concentration gradient (actually, activity gradient). Diffusion is a slow process over long distances, so the length scales over which diffusion causes composition change is typically in the range of a millimeter to centimeters.

Lastly, it is important to keep in mind that metamorphic rocks can have their compositions changed prior to metamorphism (in the classic sense). An important example is hydrothermal alteration of basalts by sea water, a common occurrence in submarine volcanic zones (Fig. 1.9D). Because the temperatures in hydrothermal systems can vary from near 0 to well over 400°C, there are a wide range of reactions that can take place. In sum, however, sea water interaction with basalts promotes: 1) growth of hydrous minerals, such as chlorite, actinolite, serpentine, and epidote, in the initially almost anhydrous rock, and 2) exchange of other chemical components between rock and sea water. Generally, at least in the modern oceans, hydrothermal alteration makes the basalt more oxidized, richer in Mg and sulfur, and poorer in Ca. One of the consequences of Mg-gain and Ca-loss is that the mafic rocks can become peraluminous, meaning that they have Al in excess of that needed to make feldspar from Ca, Na, and K. During later metamorphism (again, in the classic sense), such rocks can grow Al-rich minerals like garnet, staurolite, cordierite, and even Al-silicates, as well as a rich array of Fe-Mg amphiboles. These minerals would probably not form in rocks that had not been hydrothermally altered. In addition, the hydrothermal alteration of rocks can itself be considered a less classic metamorphic process.

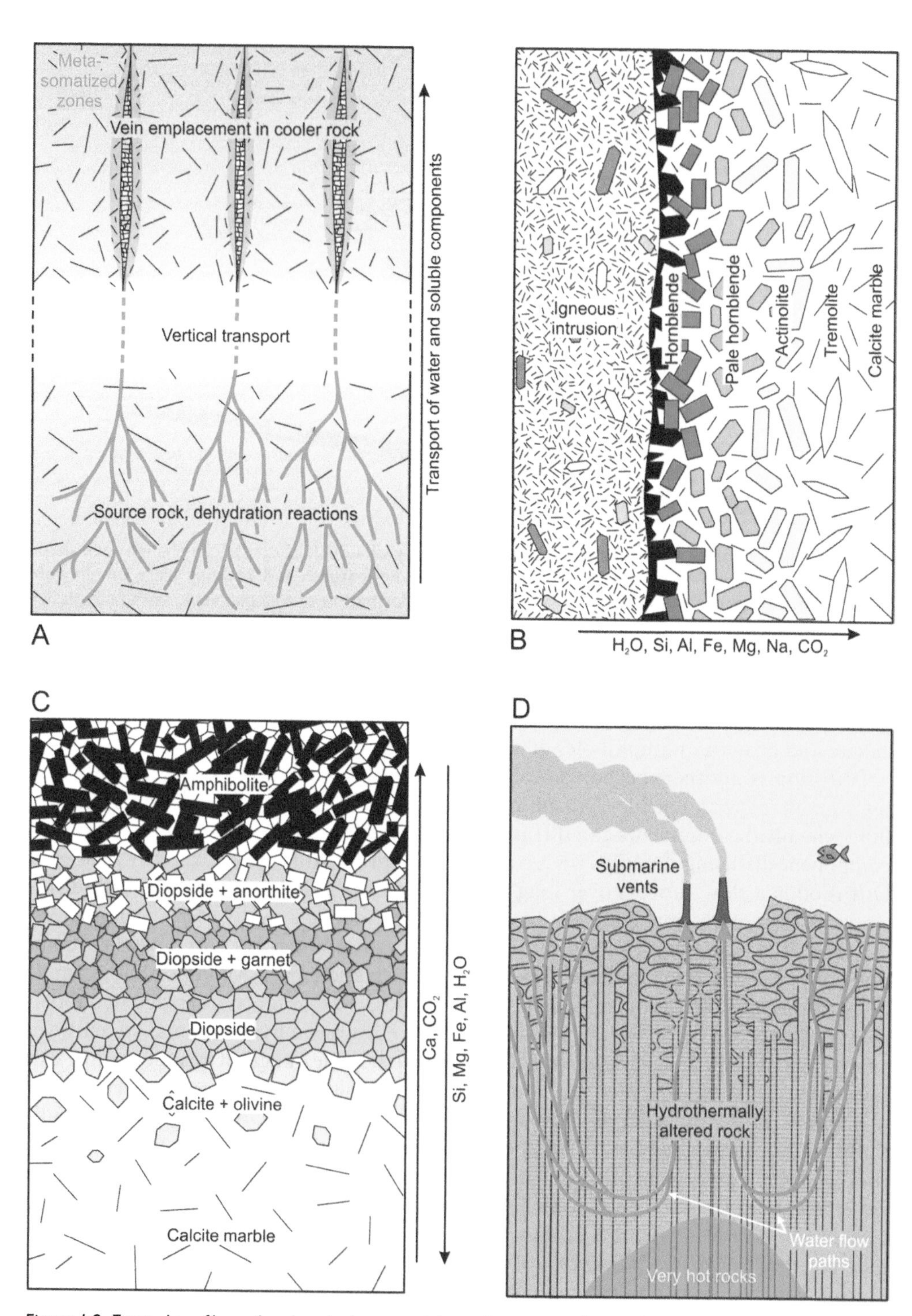

Figure 1.9 Examples of how the chemical compositions of metamorphic rocks can change. A) Dehydration reactions release water to grain boundary fluids, which migrate upwards because of their low density. The fluids carry with them dissolved materials, removing them from the source rocks. The fluids can modify cooler rocks, above, by precipitating mineralized veins and by vein margin metasomatism.

Figure 1.9 (Continued) B) Fluids migrating out of a hot intrusion bring the chemical components needed to make a range of amphibole compositions in marble. The amphiboles vary from hornblende near the intrusive contact to tremolite far from the contact. C) Diffusion of chemical components in both directions across a rock contact to produce layers of minerals and mineral assemblages that depend in part on the component diffusion rates. D) Hydrothermal metamorphism (or alteration, if you prefer) of sea floor basalts by circulating hot sea water. In addition to the mineralogical and composition changes that happen during hydrothermal activity, the changed rock compositions persist through later regional or contact metamorphism to give unusual rocks and mineral assemblages.

RECRYSTALLIZATION

There are two basic ways that rocks can deform: if they are brittle they can fracture, and if they are ductile they can flow, rather like a highly viscous fluid at scales much larger than the grain size. There is a broad transition between brittle and ductile behavior, spanning a range of pressure, temperature, rock type, and strain rate, but let us concentrate here on ductile deformation. The grain size that one sees in metamorphic rocks is the result of a number of competing processes. On the one hand, deformation tends to reduce grain size as distorted crystal lattices recrystallize into smaller subgrains (Fig. 1.10A, B). On the other hand, small grains tend to grow larger because doing so reduces the area of grain boundary surfaces, which have disordered structure and unsatisfied chemical bonds. Reducing grain surface area reduces the amount of excess surface energy, which in turn reduces the energy of the whole system. To put it another way, a few large, perfect grains are more stable than lots of little distorted ones, so the grains tend to grow (recrystallize, Figs. 1.10C, D, and E, F).

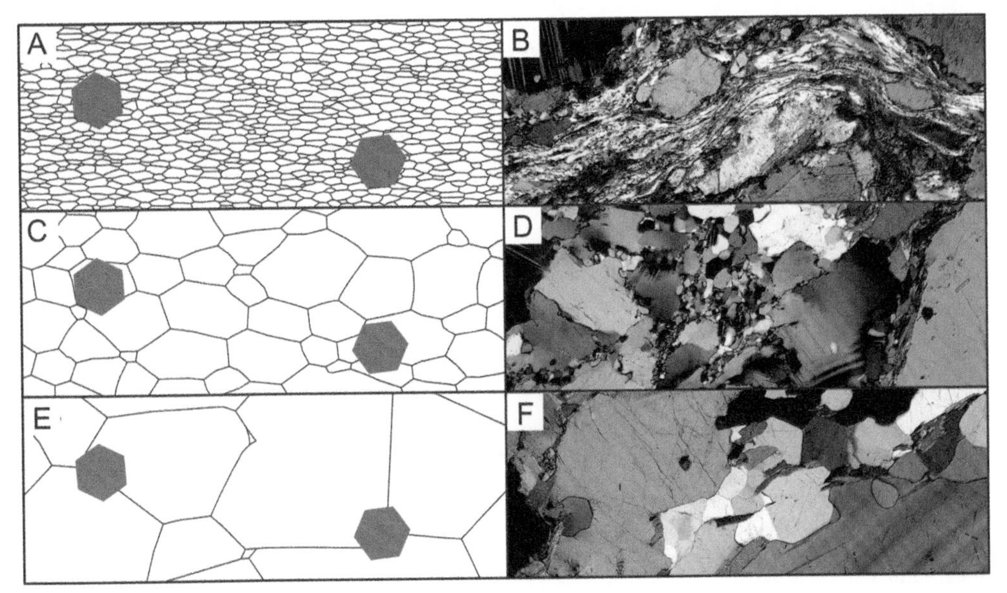

Figure 1.10 Examples of recrystallization following deformation. A, C, E) Cartoons of progressive grain growth in a mineralogically homogeneous material, such as quartzite, with two garnets as location markers. B, D, F) Thin section photomicrographs of granitic gneisses in cross-polarized light, showing progressive recrystallization in the same sequence as the cartoons. Field widths are all 4 mm. A, B) Very fine-grained, distorted biotite and quartz grains between large, relatively undeformed feldspars, with little subsequent recrystallization. C, D) Larger, more equant crystals having more regular boundaries, which have experienced some recrystallization. E, F) After considerable recrystallization, crystals become more equant and less strained, with some approximately 120° grain boundary intersections.

The grain size of a rock during deformation is therefore a result of the competition between grain size reduction driven by deformation, and grain growth driven by surface energy reduction. The rock we see at the surface may also have experienced grain growth after deformation ceased, as the rock gradually was exhumed to the surface. Once deformation stops, grain size will tend to increase, producing ever larger and crystallographically more perfect crystals with intersection angles between grain boundaries commonly approaching 120°.

Figure 1.11 shows some field examples of grain size reduction during deformation, and of grain growth and possibly erasure of some deformation fabrics during annealing. Figure 1.11A shows coarse-grained granitic gneiss (right) that has a roughly isotropic texture (same in all directions). This rock was deformed (left) causing a marked reduction in grain size, in addition to imparting a foliation. Figure 1.11B shows a strongly deformed, relatively coarse-grained gneiss (upper right). This rock was further deformed

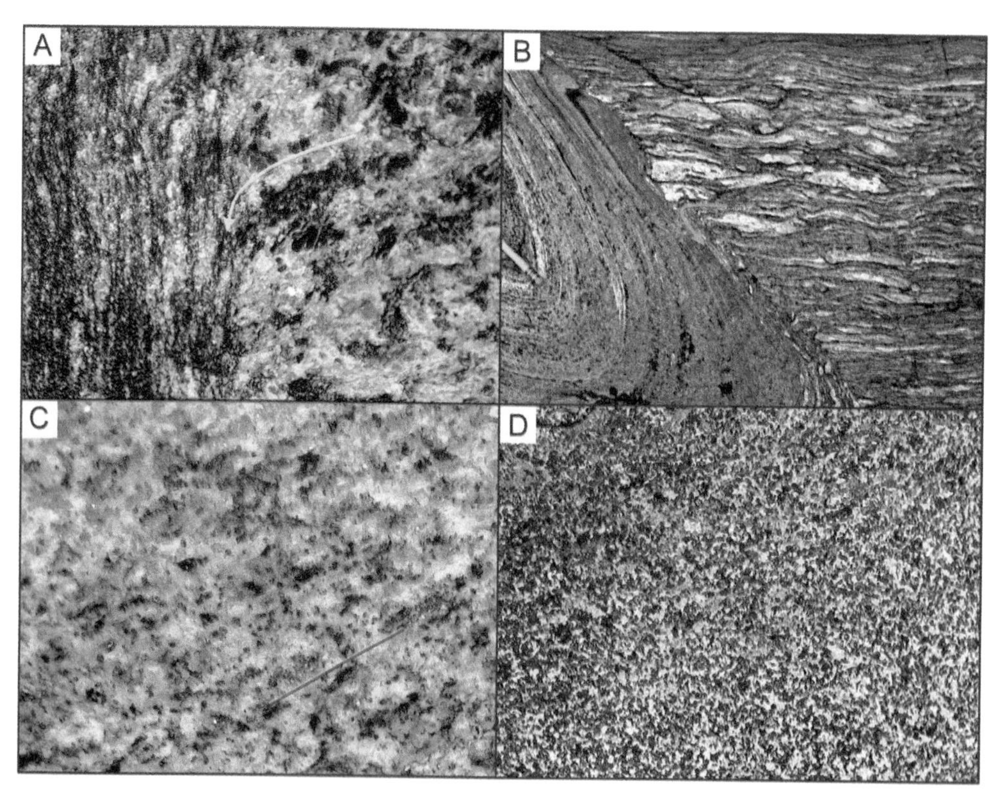

Figure 1.11 Outcrop photographs showing the effects of deformation-induced grain size reduction (A, B) and the effects of post-deformation recrystallization (annealing; C, D). A) The right side is a relatively undeformed granitic gneiss. Across a narrow boundary, the gneiss to the left has been sheared in a sinistral sense (yellow arrow, see Figs. 16.1E, F) and has a much-reduced grain size. B) To the upper right is a granitic gneiss with large feldspar augen and a horizontal foliation (blue line). This was then deformed by sinistral shear across a narrow boundary near the center of the image (blue arrow), and parallel to the pencil. To the lower left of the boundary the rock has a much finer grain size. The sheared rock has also been folded. C) Granitic gneiss containing a weak foliation parallel to the red line, defined by oriented and elongate concentrations of biotite and muscovite grains. Surrounding the micas are equant grains of quartz, feldspar, and garnet, representing partial loss of the foliated fabric by recrystallization after deformation had ceased. D) Amphibolite that is part of a layer that was strongly folded. The essentially isotropic fabric visible here resulted from recrystallization and grain growth. Widths of A, C, and D are about 10 cm.

and folded to produce a much finer-grained rock (lower left). In the boundary regions of Figures 1.11A and B, one can see the change in grain shape (indicated by arrows), the reduction in crystal size, and the increase in the number of small grains. Figure 1.11C shows a garnet-bearing granitic gneiss that was strongly deformed. Though it still retains a weak foliation of oriented biotite and muscovite crystals parallel to the red line, the fabric of minerals other than biotite is roughly isotropic, having recrystallized into roughly equant, interlocking grains of quartz and feldspar. Figure 1.11D shows an amphibolite that was also strongly deformed, judging from folds that this rock is part of. However, the amphibolite fabric is essentially isotropic with roughly equant grains. This texture is expected after a long episode of recrystallization.

SUMMARY AND INTRODUCTION TO THE FOLLOWING CHAPTERS

This first chapter was designed to introduce readers to metamorphic rocks, the processes that affect them generally, and the characteristics that most metamorphic rocks tend to have. In this sense, the introduction was for hypothetical, generic metamorphic rocks. Of course, the world is a complicated place. Metamorphic rocks come in a wide range of chemical compositions, have many different processes that affect them (more in some places than in others), and have different minerals and textures. This gives us the variety of metamorphic rock types, mineralogy, and features that we actually find in the field.

In the following chapters, the common metamorphic rock types are described and illustrated in some detail (Chapters 2-14). The rock chapters are followed by illustrations of metamorphic features, and how some of those features developed (Chapters 15-22). The last chapters illustrate protolith features that can, in some cases, survive metamorphism (Chapters 23-25). Surviving protolith features can help with the interpretation of pre-metamorphic geology. Readers will find some overlap between chapters. For example, mica schists that are discussed in Chapter 2 generally have foliations, which are discussed in detail in Chapter 15. This introduction, however, plus the Glossary, should be sufficient to ease you through the following chapters with minimal intellectual suffering.

Part I: Metamorphic rock types

Chapter 2

Pelitic rocks

The word pelite is derived from Greek, meaning clay-rock. The derivation is apt because pelitic rocks (shales etc.) are formed mostly from clays that are produced from the chemical weathering of feldspars, micas, and other minerals. Shales are the most abundant sedimentary rock type, so not surprisingly metamorphic rocks derived from them are also abundant. Although shales can themselves seem monotonous and nearly featureless, metamorphosed shales can differ strikingly from one another in appearance. They can range from extremely fine-grained slate, perhaps not so different in appearance from the original shales, to spectacular rocks bearing colorful, coarse crystals, complex textures, pegmatites, gleaming cleavage surfaces, and a host of other indicators of metamorphic history. Because Al-rich clays and quartz are commonly the dominant protolith minerals in shales, pelitic metamorphic rocks also tend to be dominated by quartz and aluminous minerals like muscovite, staurolite, the aluminosilicates, cordierite, and garnet. Pelitic rocks can host spectacular lineations, foliations, and folds at scales ranging from microscopic to kilometers across.

Metamorphosed pelitic rocks are typically gray and gray-weathering, the shade of gray depending in part on graphite content. Graphite in most cases was derived from the original sedimentary organic material present in the shale protolith. Sulfidic pelitic rocks contain abundant pyrite or pyrrhotite, usually in addition to graphite, which weather to give the rock brownish-yellow to rusty-brown surface stains. Abundant sulfides indicate an anoxic depositional environment, which speaks to marine circulation patterns and photosynthetic productivity in the original sedimentary basin. Some pelitic rocks are interlayered with quartzite, commonly derived from sandy turbidite deposits. Other pelitic rocks can be calcareous, or host calcareous horizons, possibly indicating nearby carbonate platforms supplying fine-grained carbonate detritus, or open ocean deposition above the carbonate compensation depth. Such variations can be used to distinguish different pelitic rock units from one another in the field, and so can allow mapping of them even in what might seem at first to be an endless sea of schist.

Shale deposits, especially large ones, are generally derived from weathering in and transport from large source areas. This means they were probably derived from a diverse range of source rocks, which means that, in a compositional sense, they represent an average of those rocks from a clay point of view. The result is that pelitic rocks tend to vary less in chemical composition from place to place than most rock types (if one excludes calcareous shales, at least). The similarity in chemical composition means that, as in Figure 1.2, one P-T diagram showing metamorphic fields can be broadly applicable.

The figures below most importantly illustrate textural changes in metamorphosed shales, in sequence from low to high metamorphic grades in mostly regionally metamorphosed rocks. The thin section images (Fig. 2.1) are shown to give a better sense of grain size and fine-scale foliation appearance than can generally be seen in outcrop photos. The thin sections therefore complement the field photos. The thin section photos include slate (Fig. 2.1A), phyllite (B), schist (C, D), gneissic schist that has lost all of its muscovite at high metamorphic grade (E), and a contact metamorphic granofels in which all sheet silicates have reacted away to anhydrous minerals (F). The field photos follow approximately the same sequence, from low to high metamorphic grades: Slate and other low-grade, less cleavable rocks (Figs. 2.2–2.4), phyllite (2.5, 2.6), schist (2.7–2.10), a rusty-weathering schist (2.11), and gneissic, partially melted schists at high metamorphic grade (2.12–2.15).

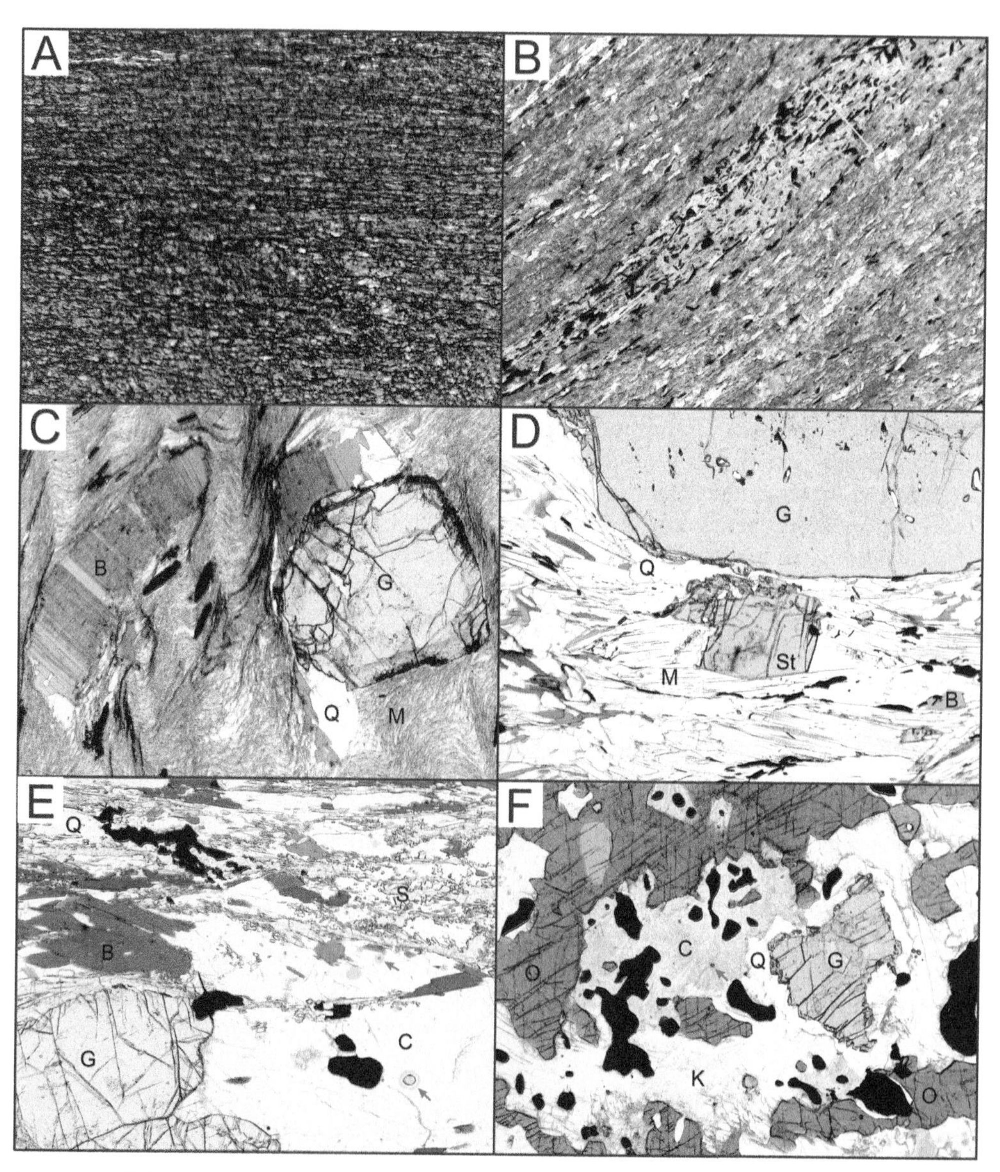

Figure 2.1 Thin section photomicrographs of pelitic rocks ranging from chlorite grade (A) to cordierite – garnet – orthopyroxene grade (F; see Fig. 1.2). These show dramatic changes in textures and mineralogy. All images are in plane polarized light, and all field widths are 4 mm. A) Slate, a

Figure 2.1 (Continued) strongly foliated rock composed mostly of chlorite, white mica, and quartz. The slate is comparatively dark in thin section because most grains are smaller than the thin section is thick, so they scatter the light more effectively than larger grains. There are also abundant, tiny grains of dark, degraded organic material or graphite. South Hero, Vermont, USA. B) Phyllite, a well-foliated rock mineralogically similar to slate (A) but with larger grains. Individual grains are not really identifiable with a hand lens, but here they show up as colorless platy white mica, platy green chlorite, colorless quartz, and dark specks of graphite and other opaque minerals. Running diagonally across the field (yellow arrow shows thickness) is a deformed fragment richer in white mica and opaque minerals than the surrounding chlorite-rich rock. Tyson, Vermont, USA. C) Schist containing the assemblage quartz – garnet – muscovite – biotite. Some of the biotite contains abundant specks of graphite, but extension during deformation fractured the biotite along its cleavage. New biotite that grew in the extensional zones is graphite-free. There is a strong foliation defined by muscovite that is folded by a nearly vertical crenulation cleavage. The red arrow points to a retrograde chlorite crystal that cuts across the foliation. New Salem, Massachusetts, USA. D) Schist containing the mineral assemblage quartz – muscovite – biotite – garnet – staurolite – plagioclase. The growth of porphyroblasts tends to distort the foliation, making it undulatory on the scale of the porphyroblasts. Pelham, Massachusetts, USA. E) Schist with the assemblage garnet – cordierite – quartz – orthoclase – sillimanite – biotite. The loss of muscovite and growth of feldspar in the upper amphibolite facies makes this rock more like a gneiss (see Chapter 8) than a schist. Sturbridge, Massachusetts, USA. F) Unfoliated granofels resulting from contact metamorphism of an aluminous gneiss against a large mafic intrusion. This rock has no remaining micas, and instead has the assemblage garnet – orthopyroxene – K-feldspar – cordierite – osumilite – quartz. Vikeså, Rogaland, Norway. Mineral abbreviations: B, biotite; C, cordierite; G, garnet; K, K-feldspar; M, muscovite; O, orthopyroxene; S, sillimanite; St, staurolite. Red arrows in E and F point to yellow alpha particle radiation halos in cordierite, surrounding uranium- and thorium-bearing mineral inclusions (probably zircon or monazite).

Figure 2.2 Typical outcrop of slate with the cleavage nearly vertical. This is a metamorphosed turbidite sequence. The quartz-rich former sandstone layers, now quartzite, are massive, without easily visible cleavage. One quartzite layer is located beneath the toe of the person's left foot, and others are indicated with yellow arrows. Some parts of the quartzites contain quartz veins that cut diagonally across the cleavage, one of which is indicated by a red arrow. These rocks, including most of the quartzites, are red-brown because they were deposited under oxidizing conditions. They contain small amounts of hematite which colors the rock. Williamsburg, Maine, USA.

Figure 2.3 Vertical sedimentary layering is clearly visible in this outcrop of metamorphosed shale and siltstone. Although slate occurs nearby, in this rock the cleavage is not so well developed. What cleavage there is cuts the nearly vertical sedimentary layering at almost right angles. The cleavage forms the flat steps that are particularly well-developed in the lower 1/3 of the image. North Carolina, USA.

Figure 2.4 Recently exposed, glacially polished surface that cuts across the rather weak slaty cleavage that is parallel to the knife. Light-colored patches are relatively coarse-grained and quartz-rich, and are interpreted to be former sandy layers (now quartzite) that were disrupted during soft-sediment deformation prior to metamorphism. Williamsburg, Maine, USA.

Figure 2.5 Interlayered foliated phyllite (darker layers) and poorly foliated quartz-rich turbidite layers (lighter, more weathering-resistant, now quartzite). The pen points in the direction of stratigraphic top, which is indicated by textural grading in the quartzite layers: coarser and quartz-rich at the sharp bottom contacts, to finer-grained and originally more clay-rich toward the gradational top contacts. Indian Pond, Maine, USA.

Figure 2.6 Muscovite – chlorite phyllite containing black, euhedral porphyroblasts of chloritoid. The rock has been broken along a prominent cleavage surface, on which there is a weak preferred mineral orientation parallel to the blue line. There is also a weak set of younger crenulations parallel to the red line (see Chapter 15). The small orange spots, most abundant in the left half of the image, are weathered ankerite crystals. Tyson, Vermont, USA.

Figure 2.7 Typical fine-grained, garnet grade pelitic schist, broken along the cleavage surface. The rock broke through some garnets, which show their red color, and around others, looking like smooth, gray bumps and pits. The matrix to the garnet porphyroblasts is made of muscovite, quartz, biotite, albite, and graphite (which gives the rock its gray color). The pen is oriented approximately parallel to a weak, younger crenulation cleavage. Charlemont, Massachusetts, USA.

Figure 2.8 Close up of kyanite – garnet – muscovite – biotite schist. In schists, mineral grains are typically large enough to see and identify by eye or with a hand lens, in contrast to slate and phyllite where the grains are too small, except for porphyroblasts. The pervasive brownish-orange color is limonite staining from weathered sulfides, which easily penetrates the muscovite cleavage and fractures around garnets. The image field width is about 8 cm. Quabbin Reservoir, Massachusetts, USA.

Figure 2.9 Typical outcrop of muscovite – biotite schist, showing one dominant foliation and cleavage surface that reflects the sunlight. This dominant cleavage has been folded on multiple scales. Big Thompson Canyon, Colorado, USA.

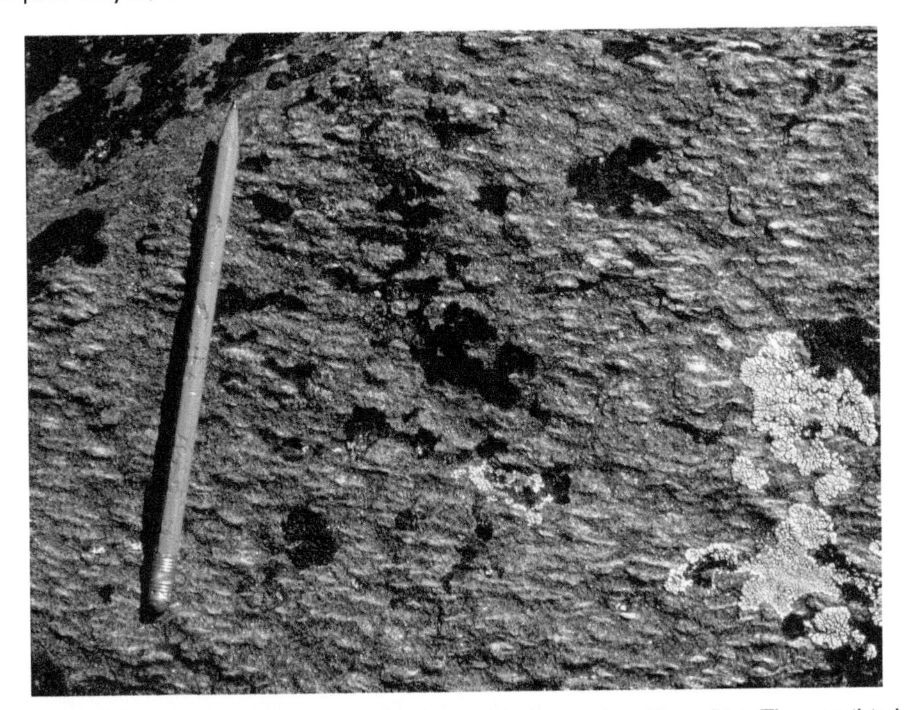

Figure 2.10 Sillimanite-rich nodules in a sillimanite – biotite – microcline schist. The pencil is lying on a flat foliation and cleavage surface, and the elongated nodules define a lineation on that surface, perpendicular to the pencil. Montezuma, Colorado, USA.

Figure 2.11 Rusty-weathering, feldspathic muscovite – biotite schist. This schist, metamorphosed to amphibolite facies, is rich in sulfides and graphite. This indicates anoxic conditions in the sedimentary basin at the time of deposition. Shown here is an outcrop surface that was a glacially polished when it was exposed by erosion in the last few decades. Weathering of sulfides releases iron, which oxidizes and precipitates on the surface as brown limonite. Sulfuric acid from the oxidation process helps to decompose the surface rock, causing it to crumble more rapidly than most other rocks. Getting fresh samples of sulfidic rock can take a lot of work. Beckett, Massachusetts, USA.

Figure 2.12 This is a somewhat inhomogeneous schist that was metamorphosed to sillimanite – K-feldspar – garnet – cordierite grade. At medium metamorphic grade this rock was interlayered muscovite schist and more calcareous biotite – quartz – plagioclase – diopside schist. At higher grade the rock partially melted, probably during muscovite dehydration, producing white pegmatite bodies of different sizes. Intense deformation after pegmatite formation caused folding and boudinage (break-up into lumps, Chapter 18) of the pegmatite bodies. Some of the schist is quite biotite-rich and dark, following melt extraction of felsic components into the pegmatite. Sturbridge, Massachusetts, USA.

Figure 2.13 Gneissic biotite schist at sillimanite – K-feldspar – garnet – cordierite grade. All of the muscovite and some of the biotite has broken down to make anhydrous phases. Lacking most of its original micas, the rock is more gneissic than typical schists at lower grade. In addition, partial melting has produced coarse leucosomes (the light colored areas, Chapter 21) of segregated, crystallized melt in which grew large garnets. The leucosomes seem to be randomly arranged, suggesting that this rock was not deformed very much after the melt crystallized. Ware, Massachusetts, USA.

Figure 2.14 Close-up of a partially melted pelitic schist metamorphosed to sillimanite – K-feldspar – garnet – cordierite grade. The rock probably partially melted during muscovite dehydration in the upper amphibolite facies, with the melt segregating into layers and blobs during deformation. The melts crystallized to form the white pegmatitic areas (leucosomes). Sillimanite is mostly difficult to see at this scale as individual crystals, though some are visible in the red ellipse and elsewhere. The common curved, anastomosing white material between the mafic minerals (yellow arrows) is almost pure, aligned, polycrystalline sillimanite. Most of the cordierite has been retrograded to an intergrowth of chlorite and muscovite (red arrows), but some is still fresh (black arrow). Wales, Massachusetts, USA.

Figure 2.15 Although the protolith for this rock was probably somewhat more feldspathic than a typical shale, this gneiss is sufficiently aluminous to have abundant garnet and minor amounts of sillimanite. During high-grade metamorphism the rock partially melted, producing coarse, white pegmatite. Later deformation disrupted the pegmatite and formed the horizontal gneissic layering. The large, white, egg-shaped object just left of center is a single microcline crystal, surrounded by recrystallized microcline. It is a remnant of the disrupted pegmatite. Comstock, New York, USA.

Chapter 3

Quartzites

Quartzites are metamorphic rocks derived from quartz-bearing sandstone or quartz-rich conglomerate, having more than 75% (NAGM, 2004) or 80% (Robertston, 1999) quartz. Sandstones are abundant in the sedimentary record and were deposited in diverse environments. They may form extensive sheets that result, after metamorphism, in regionally useful marker horizons (Fig. 3.1A, B). They may also be of more limited extent, but still form, with other rocks, parts of distinctive stratigraphic packages (Fig. 3.1C, D). Sandstones range in purity from clean beach and dune sands, to carbonate-bearing sands characteristic of some shallow ocean tropical environments, to feldspathic- or mud-rich sandstones characteristic of many river channels and offshore turbidite deposits. Quartz-rich conglomerates are more rare, and indicate a distinctive source region that could supply little but quartz rock fragments in the conglomerate size range, and a depositional environment relatively close to that source. Because of the chemical inertness, hardness, and lack of cleavage of quartz, quartzite bodies tend to weather out in raised relief above the local outcrop or landscape surface. More rarely, abundant joints can render quartzites more easily eroded than less well-jointed, tougher surrounding rocks like schist.

Because quartz itself is more or less colorless, small differences in protolith chemical composition can result in very different quartzite appearance. Quartzite color can be white or light-gray, the common colors of pure quartz, dark gray to black from admixture of graphite or more rarely tourmaline or magnetite, pink and red from hematite, and green from chlorite. Yellow, orange, and red-brown surface staining may result from the weathering of sulfides such as pyrrhotite or pyrite, carbonates such as ankerite at low metamorphic grade, or silicate minerals such as orthopyroxene in high-grade rocks. Because of surface staining, one often has to look closely at a fresh surface to see the abundant, glassy-clear, conchoidally fractured quartz grains.

Although one might think of quartzites as being featureless textural granulites (not implying pyroxene granulite facies), even where they actually do look that way they can preserve a deformational fabric in the quartz itself. Quartz c-axis alignment, caused by deformation, can be preserved and measured to aid interpretation of geologic structures. Muscovite or biotite, even if quite sparse, can define foliations, and minerals such as actinolite and sillimanite can define lineations. Deformed quartz grains or pebbles can define lineations and foliations too, but they may be subtle. Detrital minerals other than quartz, such as tourmaline and zircon, may occur, and have their own uses for provenance determination. With quartzites it is their simplicity and distinctiveness that are their charms: they are recognizable as quartzites even where deformation has thinned them from hundreds of meters to only centimeters. If the quartzite is in the same stratigraphic position, it is probably the same quartzite.

Quartzites generally have simple mineralogy, or at least the variable minerals have low abundance, so quartzites usually don't change much in outcrop appearance

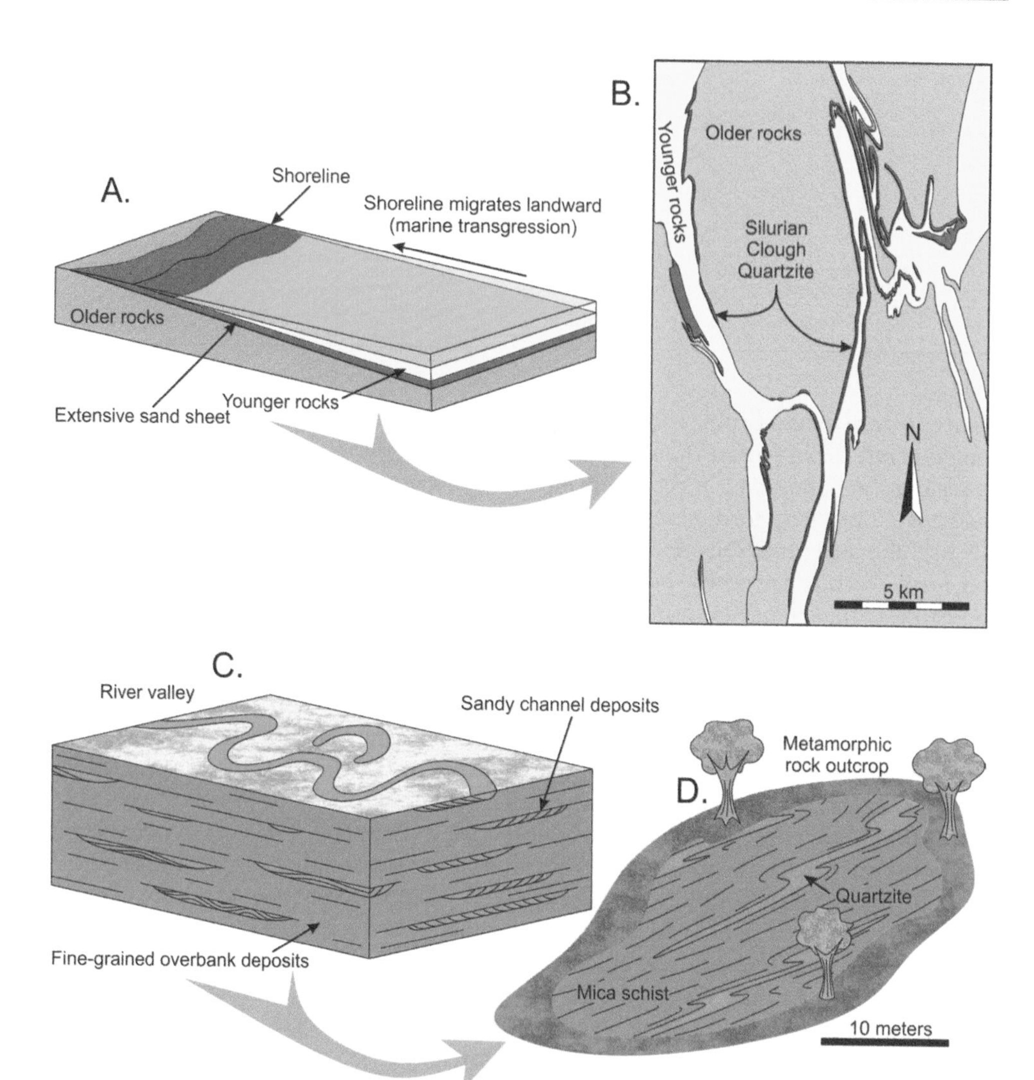

Figure 3.1 Examples of two different types of sandstone deposits that can lead to quite different outcrop patterns. Extensive sand sheets, such as coastal sands deposited during a marine transgression (A) or an extensive dune field can result in a distinctive, laterally extensive marker horizon in complexly deformed metamorphic rocks (B). This map is from north-central Massachusetts, USA (Zen, 1983), and shows the Silurian Clough Quartzite that rests unconformably on Ordovician and older rocks stratigraphically beneath it. The underlying rocks include the eroded plutonic roots of the Taconian volcanic arc that collided with Laurentia (here to the west) in the Late Ordovician. The Clough is overlain by younger Silurian and Devonian units. The Clough Quartzite is commonly a clean quartz pebble and cobble conglomerate, with clasts deformed into the shapes (and sizes) of sword blades in some places. The quartz-rich clastic materials were probably derived from erosion of the Taconian arc accretionary wedge to the west, which has abundant, easily eroded, low-grade metamorphic rocks with abundant quartz veins. In contrast, sand deposits of more restricted extent, for example river channel deposits (C), will metamorphose to form laterally discontinuous quartzite layers within other rock types (D). In this case, it will be the package of rocks that includes quartzite that will generally be mapped as a unit, rather than the quartzites themselves.

with metamorphic grade. The thin section images (Fig. 3.2) show three examples: a feldspathic quartzite that was recrystallized very little after deformation ceased, a clean quartzite with a moderate amount of recrystallization, and a mica- and garnet-bearing quartzite that is well-recrystallized. The field photos show differences in massive quartzite color and layering, and its relation to minor mineral components

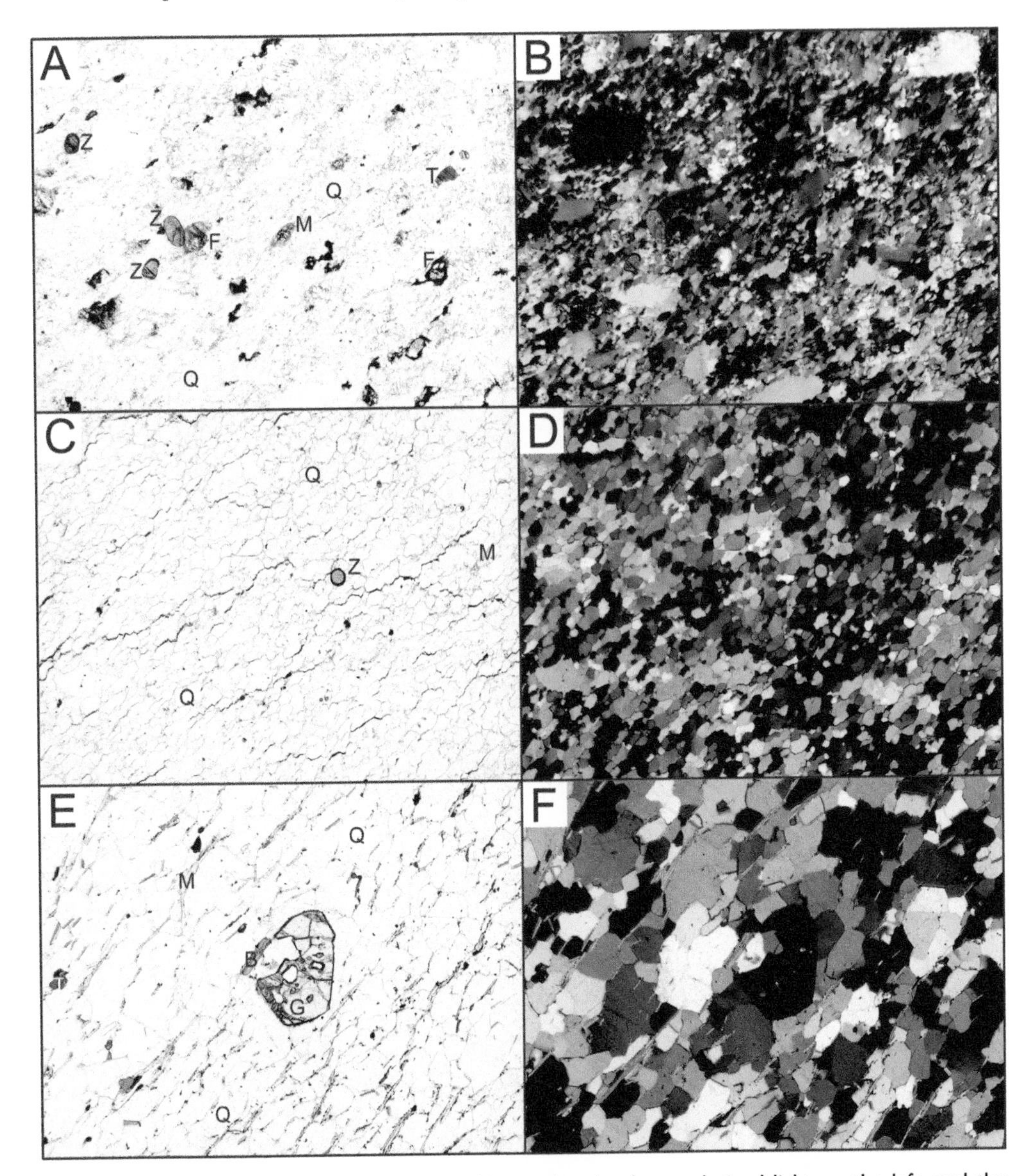

Figure 3.2 Photomicrographs of quartzite in thin section, in plane-polarized light on the left, and the same fields in cross-polarized light on the right. Field widths are all 4 mm. The images are all of strongly deformed rocks, arranged in order from having almost no recrystallization (top) to well-recrystallized (bottom). All three samples have a foliation oriented from upper right to lower left. A, B) Feldspathic quartzite, fine-grained with porphyroclasts of feldspar, quartz, and scattered detrital zircon and tourmaline crystals. The weak foliation is defined by rare muscovite and abundant, elongate quartz grains. Bennington, Vermont, USA. C, D) Partially recrystallized, very clean quartzite with only a few zircon and muscovite crystals. The weak foliation is defined by muscovite and some elongate quartz grains. Williamstown, Massachusetts, USA. E, F) Well-recrystallized garnet-bearing quartzite with the foliation defined by small amounts of muscovite and biotite. The abundance of other minerals suggests the protolith was a silty or clay-bearing sandstone. New Salem, Massachusetts, USA. Abbreviations: B, biotite; F, feldspar; G, garnet; M, muscovite; Q, quartz; T, tourmaline; Z, zircon.

(Figs. 3.3–3.5). Figures 3.6 and 3.7 show turbidite deposits, now metamorphosed to quartzite. Figures 3.8–3.10 show strongly deformed quartzites, with folds and shear zones. Figures 3.11 and 3.12 show coticule, an unusual but distinctive garnet-rich quartzite, and Figure 3.13 shows a sequence of thin quartzites that were useful for tracing complex fold structures.

Figure 3.3 Layered quartzite showing how minor differences in mineralogy, probably related to differences in the original protoliths, can result in different appearance. From left to right the layers are: 1) gray quartzite with minor graphite, 2) dark gray, brown-weathering quartzite with graphite and minor sulfides, 3) clean white quartzite, and 4) gray granitic gneiss that unconformably underlies the quartzite. Nominally colorless quartz in quartzite is easily colored by minor amounts of other minerals, and surface staining. Ørsnes, Moldefjord, Møre og Romsdal, Norway.

Figure 3.4 Because of potentially large differences in appearance, it is commonly necessary to look closely to identify the preponderance of quartz in the mineralogy. This is a feldspathic quartzite, with the feldspars appearing as tiny white spots, about 0.5 mm across, among more abundant gray, glassy quartz. Ørsnes, Moldefjord, Møre og Romsdal, Norway.

Figure 3.5 Folded, layered quartzite that has a small fault (red arrows), similar folds (most of the outcrop), and a small concentric fold near the bottom, below the pencil (see Fig. 16.1G). The concentric fold may have formed by slip along particularly muscovite-rich interlayers. There are also two white quartz veins: a folded, layer-parallel vein (blue arrow), and a crosscutting vein (black arrow). Lønset, Sør Trøndelag, Norway.

Figure 3.6 These vertical layers are composed of light-colored quartzite interlayered with darker slate. They are interpreted to have been deposited as offshore, marine turbidites and shale. The original turbidite deposits were graded with respect to grain size, with the lower parts of each layer, to the right, being sand-rich, progressively grading to finer-grained, more silty or clay-rich bed tops to the left (red arrow). Though the apparent bedding may be somewhat transposed, the grading remains clear. Wells, Maine, USA.

Figure 3.7 Graded quartzite layer, interpreted to be a metamorphosed turbidite deposit, in between layers of sillimanite schist. Stratigraphic top is toward the top of this image, from the sharp lower

Figure 3.7 (Continued) contact of the quartzite to the gradational upper contact. The schist above and below the quartzite is much coarser-grained than the quartzite itself because of growth during metamorphism of large mica and garnet crystals, and large andalusite porphyroblasts, now pseudomorphed by sillimanite. Jaffrey, New Hampshire, USA.

Figure 3.8 Folded muscovite- and graphite-bearing, thinly layered quartzite. In highly deformed rocks such as these, primary sedimentary structures are difficult to discern. Note the contrast between the gray quartzite in most of the image, and the white quartz vein in the upper right. Skår, Moldefjord, Møre og Romsdal, Norway.

Figure 3.9 Gray, complexly folded quartzite, with layers that are alternately rich and poor in muscovite and kyanite. Contrasting with the gray quartzite are abundant, deformed, white quartz veins. Many of the quartz veins have been boudinaged into white lumps. Black Mountain, New Hampshire, USA.

Figure 3.10 Sandstones commonly contain primary depositional features, but those are usually highly modified or destroyed during metamorphism. The strongly deformed layers shown here may once have been sedimentary bedding, but are now transposed (see Chapter 15). What appears to be cross-bedding has actually been produced by a shear zone along the surface between the two red arrows. The layers above the shear zone have been thinned along it, and rotated with respect to the layers below the shear zone. Vestnes, Moldefjord, Møre og Romsdal, Norway.

Figure 3.11 A rare but striking rock type found in some metamorphosed sedimentary and volcanic successions is pink, garnet-bearing quartzite known as coticule. This is a thin, doubly plunging, folded layer of coticule in pelitic schist. Coticule is generally interpreted to have been derived from beds of impure, Mn- or Fe-rich chert. In contrast to host rocks that may have millimeter-scale garnets, the garnets in coticule tend to be very small, commonly less than 20 μm in diameter, and very numerous. Though rare, these rocks can help form useful marker horizons. Jaffrey, New Hampshire, USA.

Figure 3.12 Close-up of a coticule layer near the one in Figure 3.11. Garnets in the enclosing schist are 3-6 mm in diameter, but are much smaller and more abundant in the coticule. Jaffrey, New Hampshire, USA.

Figure 3.13 Even where they are not dominating the landscape, quartzite layer sequences can help define structures in otherwise monotonous rocks. In the image center are several white quartzite layers, originally turbidite deposits, within coarse, gray pelitic schist. This concentration of laterally continuous quartzite layers helped define isoclinal folds and other structures during mapping (Thompson, 1988). Jaffrey, New Hampshire, USA.

Chapter 4

Marbles

Marble forms from metamorphism of carbonate-rich sedimentary rocks, such as limestone or dolostone. To be classified as marble the rock must have more than 50% carbonate minerals (NAGM, 2004). Marbles can vary widely in appearance, depending on their original composition and sedimentary layering, metamorphic conditions, the amount of deformation, and the amount of post-deformation annealing.

Limestones originally containing only calcite, or calcite and quartz (as sand, for example), tend to change mineralogy very little except at the highest metamorphic temperatures, or at very low pressures such as during contact metamorphism. Dolomitic, quartz-bearing marbles tend to be more reactive. Decarbonation of carbonates frees up components such as MgO and CaO that react readily with quartz to make diopside, olivine, and other minerals, for example:

$$\underset{\text{dolomite}}{CaMg(CO_3)_2} + \underset{\text{quartz}}{2SiO_2} = \underset{\text{diopside}}{CaMgSi_2O_6} + \underset{\text{fluid.}}{2CO_2} \quad (4.1)$$

The reactions that take place depend on the local fluid composition in addition to pressure and temperature. CO_2-rich metamorphic fluids tend to stabilize carbonate minerals to higher temperatures and lower pressures, whereas H_2O-rich fluid compositions tend to destabilize carbonates and stabilize hydrous minerals.

In addition to quartz, impure limestones commonly contain clays, iron oxides, and other minor components that are sources of Al, Na, K, Fe, and Ti. Assuming metamorphic fluids have intermediate CO_2-H_2O proportions, these can result in metamorphic assemblages that include grossular, diopside, tremolite, phlogopite, spinel, magnetite, Ca-rich plagioclase, titanite, scapolite, K-feldspar, and a host of others. Figure 4.1 shows a phase diagram for one marble composition, schematically showing the characteristic progression of mineralogy for impure marbles: hydrous minerals and dolomite at low temperatures are replaced by anhydrous minerals and loss of dolomite at high temperatures, aragonite occurs only at high pressure and low temperature, and wollastonite appears only at relatively low pressure and high temperature. Many limestones also contain some organic matter, which becomes graphite during metamorphism.

Rocks having less than 50% carbonate minerals are usually classified according to their other mineral or textural components. For example, if the rock is dominated by Ca-rich silicates such as diopside, grossular, epidote, and tremolite, it would probably be classified as a calc-silicate rock, or scarn. If instead the rock is mica-rich and foliated, it might be called a calcareous schist. These are considered in Chapters 5 and 6, respectively.

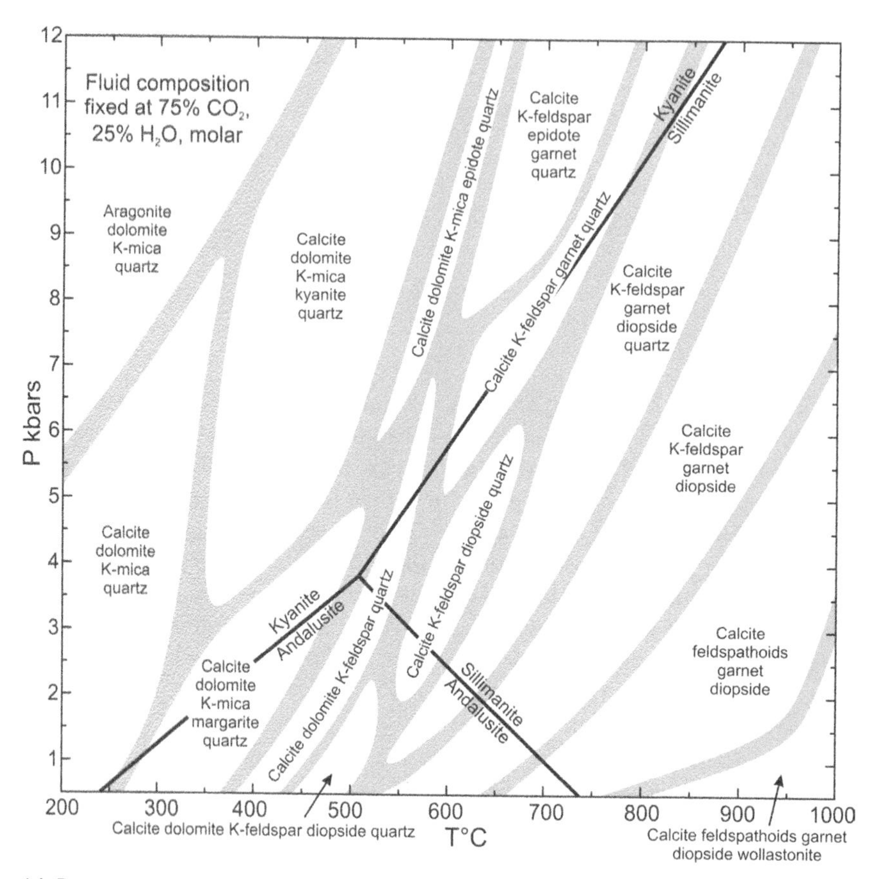

Figure 4.1 Pressure-temperature phase diagram calculated for one particular marble composition. The main points of interest are: 1) Calcite is replaced by aragonite at high pressure and low temperature. 2) K-mica is the principal host for K_2O at low temperatures, replaced by K-feldspar at intermediate temperature, and by feldspathoids at high temperature (feldspathoids are feldspar-like minerals such as nepheline and kalsilite). 3) Dolomite is lost at high temperature both by dissolving into calcite and by decarbonation and reaction with silica and other chemical components to produce silicates. 4) At yet higher temperatures quartz is lost to produce more silicates, including wollastonite at low pressure. This diagram is shown only as an example, and should not be used to estimate the pressures and temperatures of formation for marbles generally. Unlike the more widely applicable facies or grade diagrams for basaltic or pelitic rocks (Figs. 1.1 and 1.2), marbles vary enormously in composition, so no one diagram like this has much significance for marbles of very different composition. This phase diagram was calculated using the thermodynamic modeling software Perple_X (Connolly, 2009). The marble chemical composition used was SiO_2, 12.45%; Al_2O_3, 2.81%, Fe_2O_3, 0.80%; FeO, 0.84%; MgO, 3.30%; CaO, 79.06%, and K_2O, 0.75% (average of eight limestones, Wheeler, 1999). Cpx is the clinopyroxene diopside, and garnet is a grossular-andradite-rich solid solution.

The thin section photographs show marbles at three metamorphic grades in sequence from greenschist facies to pyroxene granulite facies (Fig. 4.2). The field photos show marbles from greenschist (Fig. 4.3), epidote amphibolite (Figs. 4.4–4.7), amphibolite (Fig. 4.8), and granulite facies (Figs. 4.9–4.13). The last two field photos show some special features that result from marble dissolving at Earth's surface (Figs. 4.14, 4.15). Marble is quite water-soluble and can contain a number of readily accessible chemical elements needed by living things (e.g., Ca, Mg, from carbonates, P from apatite, K from micas), marble outcrops tend to become rapidly covered with lichens, moss, and so on. Take heart, though, it only takes a few square centimeters of exposure to see if it fizzes under a drop of dilute acid.

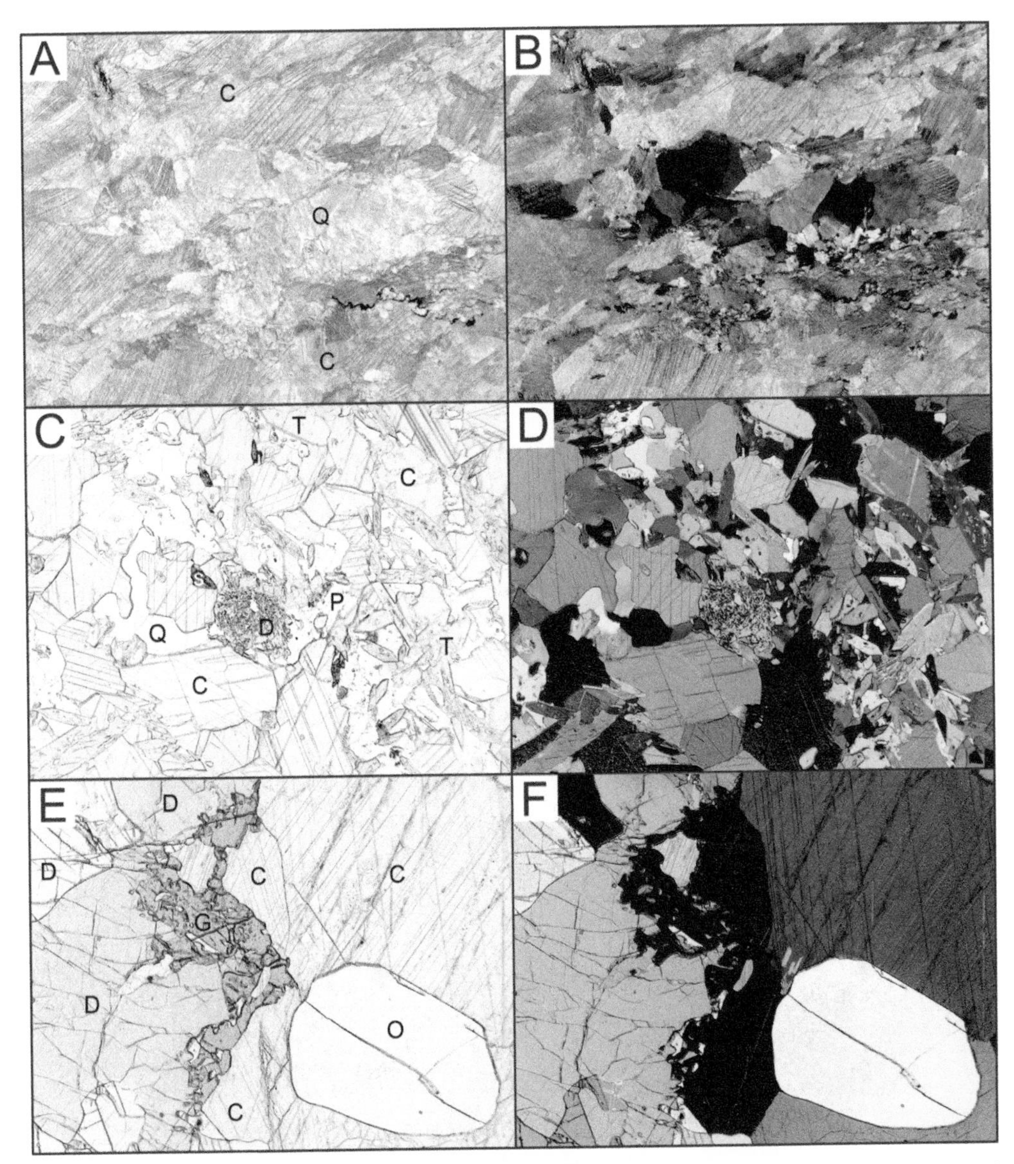

Figure 4.2 Thin section photomicrographs of marbles. All images on the left are in plane-polarized light, and those on the right are of the same fields in cross-polarized light. Field widths are all 4 mm. The marbles are shown in order of metamorphic grade: lower greenschist facies (A, B), amphibolite facies (C, D), and granulite facies (E, F). A, B) Quartz-bearing marble associated with slate, a small fragment of which is the dark material to the lower right. The quartz, possibly originating as a now disrupted quartz vein, is full of fluid inclusions along healed fractures, making it somewhat cloudy. With optical microscopy, quartz and calcite are the only identifiable minerals present. Other minerals, doubtless present in the slate fragments, are too small to identify. South Hero, Vermont, USA. C, D) Marble containing about 50% calcite in addition to diopside, tremolite, plagioclase, quartz, and titanite. In D, some of the plagioclase can be seen to be zoned (red arrow points out the best example). Gassetts, Vermont, USA. E, F) Silicate-rich area in marble, containing the assemblage calcite – diopside – grossular-andradite garnet – olivine. Cascade Lake, Keene, New York, USA. In all sections the calcite has abundant cleavage cracks, and deformation twins that can give rise to colorful stripes. Abbreviations: C, calcite; D, diopside; G, grossular-andradite garnet; P, calcic plagioclase; Q, quartz; S, titanite; T, tremolite.

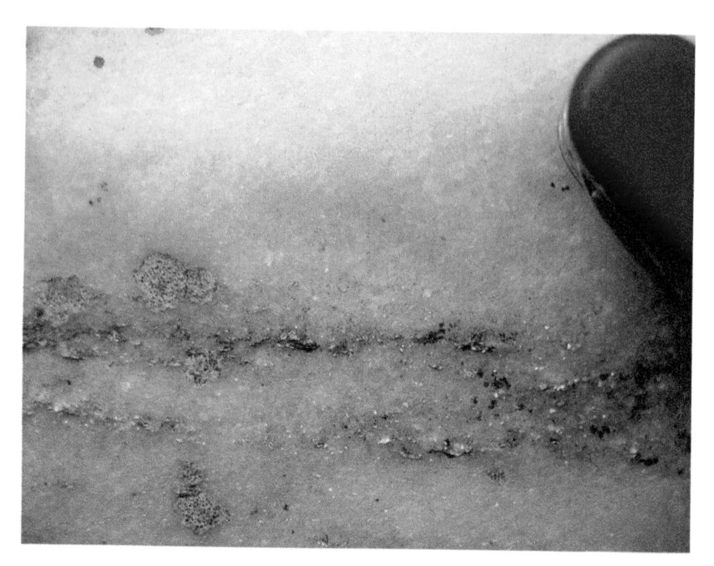

Figure 4.3 Close-up of marble metamorphosed to upper greenschist facies. Post-deformation recrystallization and grain growth is indicated by the abundant, equant calcite grains that are 0.5–1 mm across. In the lower half of the image are layers rich in phlogopite (nearly colorless so it looks like muscovite). North Adams, Massachusetts, USA.

Figure 4.4 Clean marble (right) in contact with garnet – biotite schist. As a result of dissolution by slightly acidic rain, marble in wet climates tends to weather faster than silicate rocks. In this case the marble has not been strongly weathered because it is a recently exposed, glacially polished surface. The outcrop surface extends relatively smoothly from marble to the schist. The schist, a more brittle rock, was fractured near the contact long before they were exposed at the surface. Kvithylla, outer Trondheimsfjord, Sør Trøndelag, Norway.

Figure 4.5 Typical wet climate weathering surface on gray marble. The crystalline marble weathers in cuspate hollows that result from dissolution by mildly acidic rainwater. The individual calcite grains, about 1–2 mm across, are visible. Bolsøy, Moldefjord, Møre og Romsdal, Norway.

Figure 4.6 The brownish-gray rock to the left and right is brown-weathering, impure marble. The marble is exposed on either side of an amphibolite layer, between the yellow arrows, that is covered with yellow and light-gray lichen. The rocks are folded about an antiformal hinge that plunges gently downward into the water. Bolsøy, Moldefjord, Møre og Romsdal, Norway.

Figure 4.7 Shoreline outcrop of strongly folded, alternating layers of carbonate-bearing biotite schist and white to gray marble. The schist weathers more slowly, so it stands in raised relief above the marble surfaces. The gray coloring in some of the marble layers is caused by small quantities of graphite. Vikan, outer Trondheimsfjord, Sør Trøndelag, Norway.

Figure 4.8 Close-up of marble containing angular, dark-gray amphibolite fragments (upper and lower right). The black mineral in the center is tourmaline, and the rest is calcite, silicate minerals including quartz and diopside, and graphite. This is a close-up of the marble-hosted breccia shown in Figure 7.15. Paradox, Adirondacks, New York, USA.

Figure 4.9 Coarse-grained, light-gray marble with folded dark-brown calc-silicate layers snaking through it. The calcite grain size in most of the marble is about 1 cm, but up to 10 cm in some places (bottom center). This outcrop is covered by the Hudson River during high water from early spring snow melt runoff. Because of ice and bed load scouring action, the outcrop is kept largely free of lichen. Warrensburg, New York, USA.

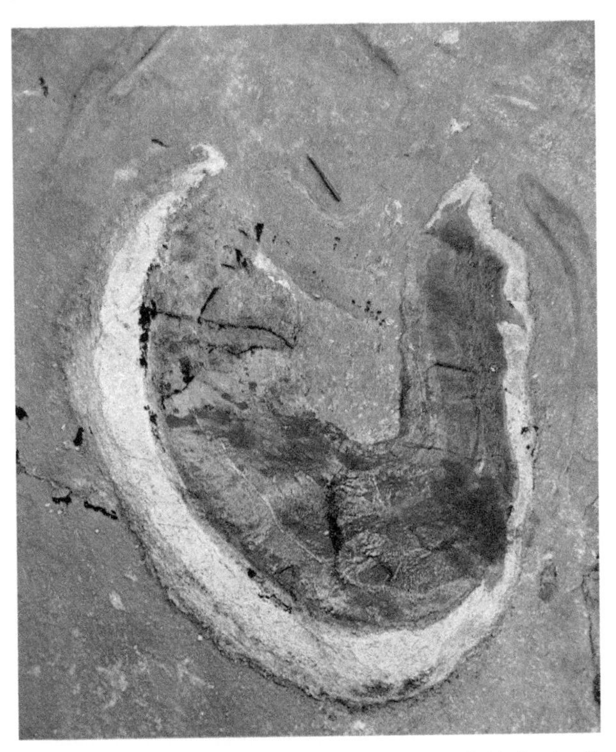

Figure 4.10 Marble enclosing a detached fold hinge amidst thinner, folded calc-silicate layers and layer fragments. The large fold hinge has dark amphibolite and calc-silicate rock on the inside, with a white granitic gneiss layer on the outside. Because of the enormous amount of strain that highly ductile marbles can experience, silicate rocks within them can become greatly thinned, to the extent that the thicker fold hinges become detached from their limbs. With isolated fold hinges, stretched and fragmented fold limbs, boudinaged calc-silicate layers, and fragments broken off the contacts, marbles can become, in effect, tectonic breccias containing a wide range of rock types. Warrensburg, New York, USA.

Figure 4.11 Close up of a granulite facies marble, in which recrystallization has erased most small-scale structural fabric. The assemblage is calcite (white) – diopside (dark-green) – forsterite (light green). Light-brown spots in the upper 1/3 of the image are serpentinized olivine (two indicated by red arrows), which formed when fractures allowed water access to some olivine crystals. Keene, Adirondacks, New York, USA.

Figure 4.12 This is an example of the ductility of marbles under high-grade metamorphic conditions. The gray rock is a garnet-bearing charnockitic gneiss (charnockite is an orthopyroxene-bearing granite). The white dike cutting through the charnockite is marble, 10–20 cm thick. The marble is interpreted to have flowed in the solid state into an opening fracture in the gneiss. This marble dike has the same texture, mineralogy, and mineral proportions as large marble bodies nearby, so it is unlikely to be a fluid-deposited vein or a carbonatite dike. The marble contains numerous angular fragments of the host gneiss as xenolith-like breccia fragments. A large one is visible in the center of the photograph (red arrow). Schroon Lake, New York, USA.

Figure 4.13 A close-up of another apparent marble dike that was injected into a fracture between amphibolite, below, and garnet-rich granitic gneiss, above. The marble contains nearly black diopside and graphite. The original marble was probably relatively fine-grained, but recrystallization appears to have transformed those grains into a few very large single crystals, as indicated by regions having the same calcite crystal cleavage orientations. Ticonderoga, Adirondacks, New York, USA.

Figure 4.14 Because of the high solubility of carbonates in water, silicate layers within marble can weather out in spectacular raised relief to give a three-dimensional view of the structures. Unfortunately, because of the large ductility contrast between marble and silicate rocks, folded layers in marble may not reliably represent larger scale surrounding structures. Fjørtoft, Nordøyane, Møre og Romsdal, Norway.

Figure 4.15 In wet climates marbles tend to erode faster than adjacent silicate rocks. Long grooves in the outcrop surface, or valleys in the landscape, may hint at marble even where it is covered with lichen, grass, soil, and swamps. Here, a thin marble layer, between the yellow arrows, is sandwiched between layers of garnet – biotite schist. Dissolution has lowered the marble surface below that of the surrounding silicate rocks. Vardholmbukta, Moldefjord, Møre og Romsdal, Norway.

Chapter 5

Calc-silicate rocks

Calc-silicate rocks, also called skarn, are composed mostly of calcium-rich silicate minerals such as diopside, anorthite, grossular-andradite garnet, and wollastonite, though they may contain some calcite, quartz, and other, non-calc-silicate minerals. Calc-silicates can form directly from calcareous sedimentary or volcanic rocks (e.g., calcareous shale, carbonate mud mixed with reworked pyroclastics), in hydrothermal veins, at igneous contacts where plutons crosscut carbonate rocks (Fig. 1.9B), and at contacts between marble and silicate rock (Fig. 1.9C). Calc-silicate rocks associated with pluton contacts can form ore deposits of economic importance. Formation of calc-silicate rocks at rock contacts commonly involves exchange of chemical components, generally with marble supplying the Ca and possibly Mg, released from carbonates during decarbonation reactions, and the silicate rock supplying silica and possibly other components. Reactions between these components in the contact zone produce the calc-silicate minerals.

Although calc-silicate rocks are rarely extensive enough to form their own geologic units, their characteristic occurrence as parts of some stratigraphic successions, for example in some pelitic or volcanic sequences, can be a distinctive characteristic for identifying different units. Calc-silicate rocks in an extensive unit may indicate a near-shore carbonate shelf-type environment, or possibly slow clastic sedimentation in a marine environment above the carbonate compensation depth.

The sequence of photos below is two-fold. The thin section photos (Fig. 5.1) are arranged from low-grade to high-grade, to a certain extent showing mineralogical changes with grade. Because of wide variations in chemical composition, the reader is cautioned that mineralogy, in these rocks, is not always a reliable guide to metamorphic grade in the sense of Figures 1.1 and 1.2. The field photos are arranged first by origin, and secondarily by increasing metamorphic grade, where possible. Figures 5.2–5.4 show calc-silicate boudins that are interpreted to have once been stratigraphic layers, indicated by their presence in stratigraphic successions of metamorphosed sedimentary or volcanic rocks. Figures 5.5 and 5.6 are interpreted to have been hydrothermal veins, based on their occurrence in metamorphosed plutonic rocks. Figures 5.7 and 5.8 are also interpreted to have formed as veins, because of their stark mineralogical contrast and crosscutting relationships with their host rocks. Figures 5.9–5.11 are products of reactions between silicate rock and marble, or contact metamorphism. Figure 5.12 is an interesting example of a calc-silicate outcrop puzzle.

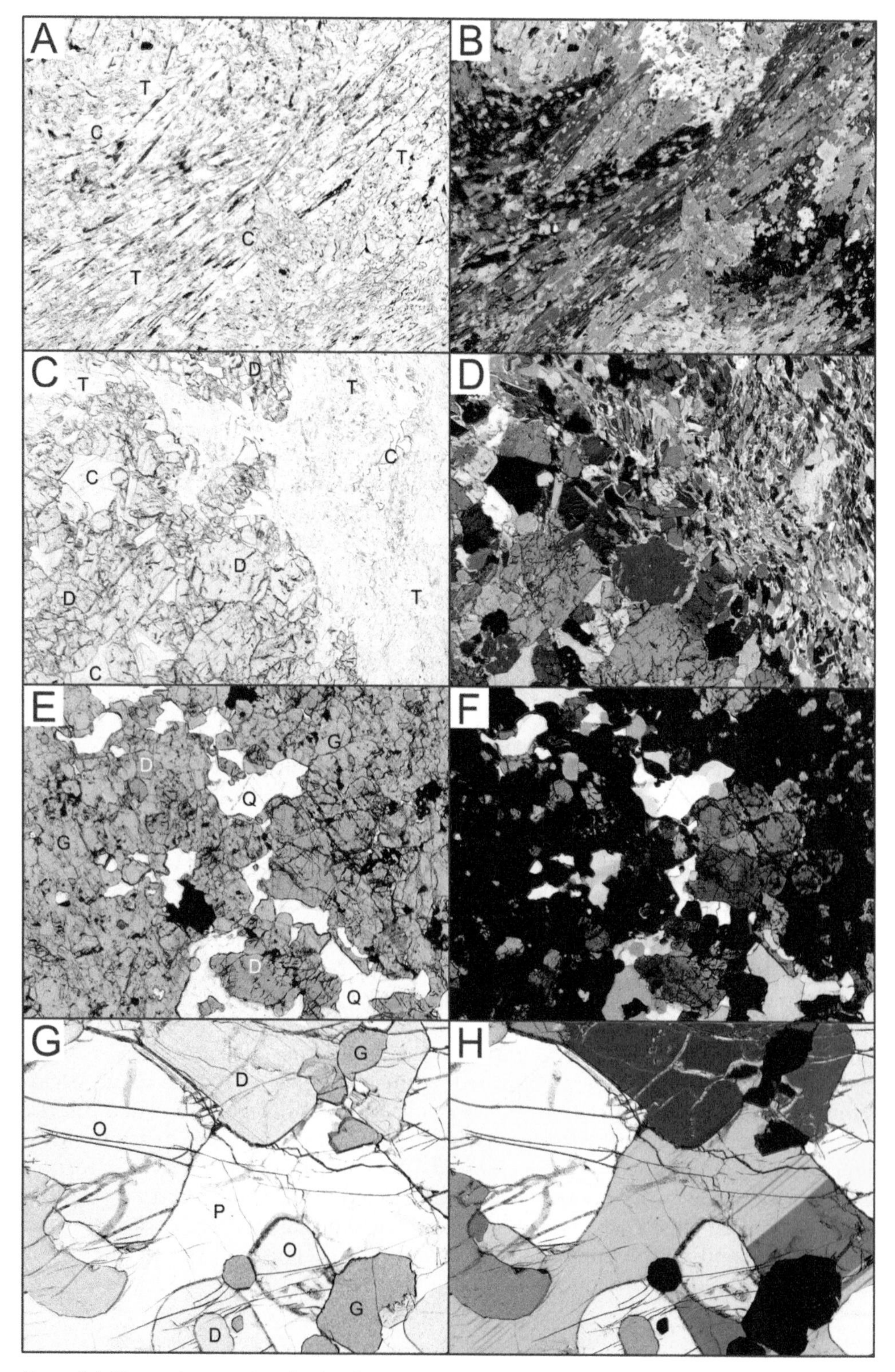

Figure 5.1 Photomicrographs of calc-silicate rocks at different metamorphic grades. Images on the left are in plane-polarized light, and those on the right are the same fields in cross-polarized light.

Figure 5.1 (Continued) All field widths are 4 mm. A, B) Calc-silicate rock at greenschist facies, composed of tremolite and calcite. The black material is graphite that was mostly pushed aside by growing tremolite grains. Bath, New Hampshire, USA. C, D) Calc-silicate rock at lower amphibolite facies, composed of diopside, tremolite, and calcite. Deformation was accommodated primarily in the tremolite-rich parts, indicated on the right half of D by the preferred tremolite grain orientations. Bolton, Massachusetts, USA. E, F) Oxidized, Fe-rich calc-silicate rock at upper amphibolite facies, composed of andradite-rich garnet, diopside, quartz, and magnetite. This is part of a probable fossil hydrothermal plumbing system in mafic volcanic rocks. Within centimeters to meters of this rock are different layers that are variably rich in magnetite, garnet, diopside, and sulfides. Interlayered with them were amphibolites that contain abundant Fe-Mg amphiboles (cummingtonite, anthophyllite, gedrite) that are indicative of unusually low whole rock Ca/Mg ratios. Such low ratios can result from hydrothermal alteration of normal basaltic rocks by high-temperature sea water (Schumacher, 1988). Quabbin Reservoir, Massachusetts, USA. G, H) Calc-silicate rock at granulite facies, composed of diopside, olivine, plagioclase, and grossular-andradite garnet. Cascade Lake, Keene, New York, USA. Abbreviations: C, calcite; D, diopside; G, grossular-andradite garnet; O, olivine; P, plagioclase; Q, quartz; T, tremolite.

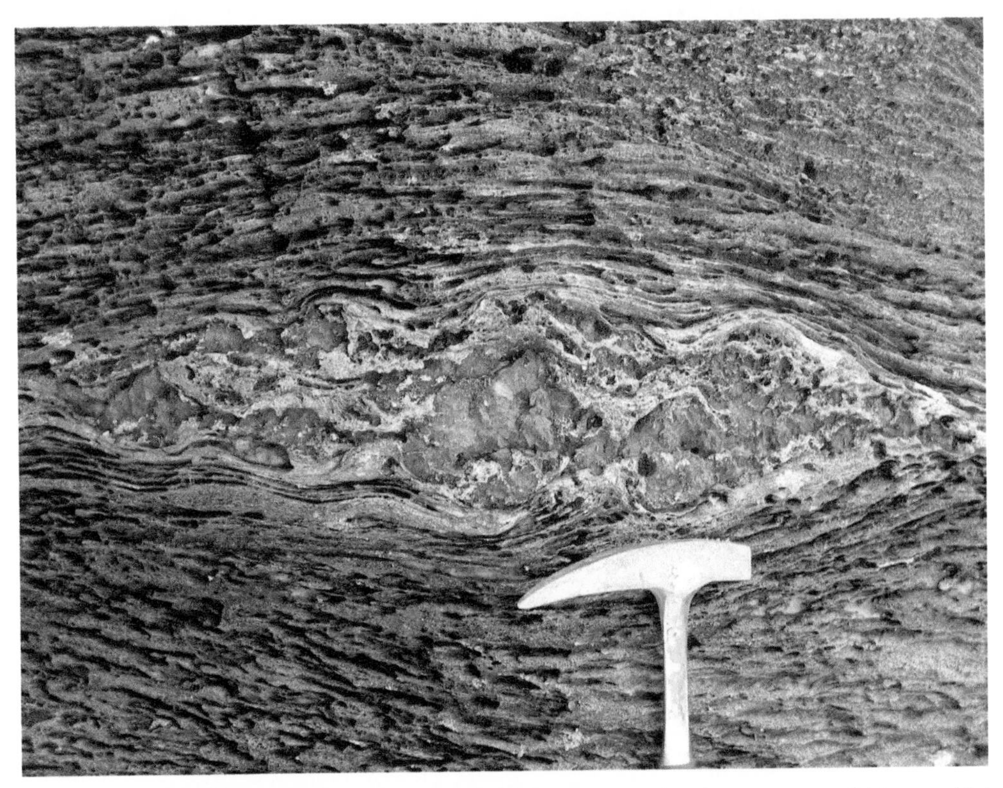

Figure 5.2 Carbonate-calc-silicate boudin in layered, calcareous biotite – quartz – feldspar schist at epidote amphibolite facies. The carbonate portion of this boudin is brown, in the center of the image. This example shows a rim around the carbonate of light-gray calc-silicates that include green epidote and nearly colorless actinolite and diopside. Kvithylla, outer Trondheimsfjord, Sør Trøndelag, Norway.

Figure 5.3 Calc-silicate boudin in amphibolite at upper amphibolite facies. The inhomogeneous boudin center is made of red grossular-andradite garnet, dark-green diopside, and white anorthite. The outer parts are mostly hornblende, diopside, and calcic plagioclase. This was probably once part of a more continuous layer. Quabbin Reservoir, Massachusetts, USA.

Figure 5.4 Boudins made of interlayered calc-silicate and amphibolite in migmatitic gneiss, metamorphosed at lower granulite facies. The host rocks are garnet – K-feldspar – Al-silicate ± cordierite gneisses that melted, producing a highly heterogeneous and complexly deformed migmatite. The more mafic blocks are composed mostly of diopside, plagioclase, and hornblende, with minor

Figure 5.4 (Continued) amounts of grossular-andradite garnet in the lighter-colored diopside-rich layers. The proportions of minerals in the mafic rocks varies considerably, but the diopside-rich layers are too calcareous to have had a purely volcanic protolith. These were probably interlayered dolomitic limestone and mafic volcanics, between thick shale beds. During metamorphism, the carbonates reacted with adjacent amphibolite to produce the diopside-rich layers. Åre, Jämtland, Sweden.

Figure 5.5 The host rock is strongly deformed tonalitic gneiss and black amphibolite. The greenish boudins across the middle of the image are a calc-silicate rock made of epidote with minor chlorite, titanite, quartz, and calcite. These rocks have been metamorphosed to upper epidote amphibolite facies conditions. The calc-silicate rock is interpreted to have originally been a hydrothermal vein that cut the adjacent plutonic rocks at some time after host rock solidification. Ductile deformation took place long after the rocks themselves were emplaced. Almvikneset, outer Trondheimsfjord, Sør Trøndelag, Norway.

Figure 5.6 Calc-silicate boudin in inhomogeneous tonalitic gneiss, metamorphosed to amphibolite facies conditions. The calc-silicate is composed mostly of epidote, diopside, and quartz. It is thought

Figure 5.6 (Continued) to be a disrupted hydrothermal vein that post-dated pluton solidification, but long pre-dated regional metamorphism and deformation. Different sized, mineralogically identical boudins are strung out in a line to the right of this photo. Quabbin Reservoir, Massachusetts, USA.

Figure 5.7 Boudinaged quartz vein (yellow arrow) and later crosscutting calc-silicate vein (red arrow) in sillimanite – muscovite schist. This calc-silicate vein was precipitated by Ca-rich hydrothermal fluids that flowed through a late-opening fracture. The white vein margins are made of quartz, calcite, and anorthite, and the pink core is made of grossular-andradite garnet, diopside, and minor quartz and anorthite. Mount Moosilauke, New Hampshire, USA.

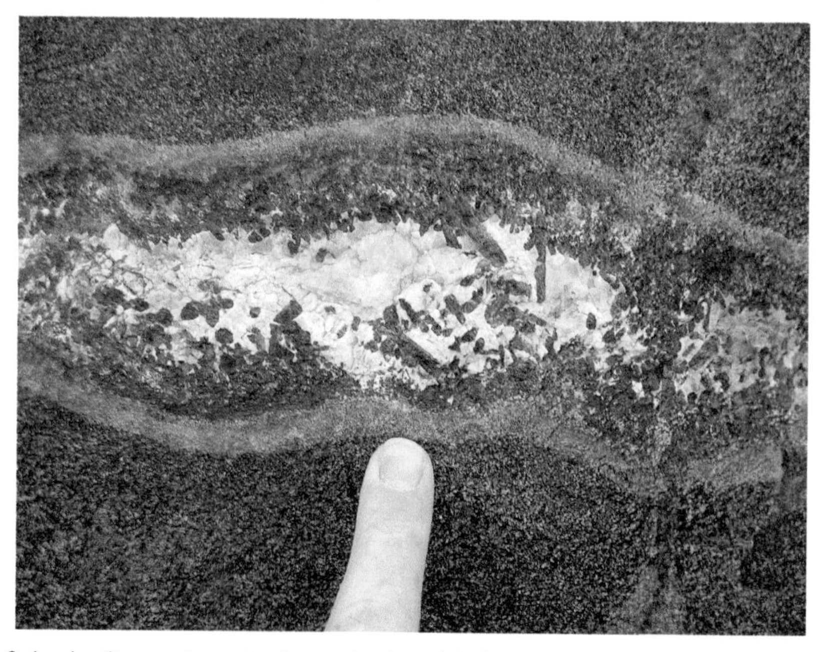

Figure 5.8 A calc-silicate vein cutting fine-grained amphibolite at upper amphibolite facies. The vein is zoned with a rim made of black to dark-green, inward-coarsening, euhedral diopside crystals, and a

Figure 5.8 (Continued) white center made of anorthite and calcite. The brownish margin at the finger tip, and on the opposite side of the vein, is Ca-metasomatized amphibolite, now made mostly of diopside, anorthite, and titanite. Warrensburg, Adirondacks, New York, USA.

Figure 5.9 Calc-silicate contact zone between marble (dissolved away, used to be in the region above the yellow dotted line) and granitic gneiss (lower half of the image). In a traverse from the brownish-gray gneiss, the layers are: 1) calcic plagioclase – K-feldspar – diopside – quartz, 2) anorthite – diopside – quartz, and 3) garnet – diopside – anorthite, with the garnet being a grossular-andradite solid solution. The layers were produced by double-diffusive transport of different chemical components across the contact, principally silica moving toward the marble and calcium toward the gneiss. Warrensburg, New York, USA.

Figure 5.10 This is a coarse-grained marble that has engulfed large blocks of granitic gneiss (top and bottom right), that are surrounded by black selvages of diopside – anorthite – grossular-andradite – graphite calc-silicate rock. The marble also contains a long, folded ribbon of rusty-weathering diopside – olivine calc-silicate rock. The ribbon may be the remains of a severely extended and thinned fold limb. Weathering of pyrrhotite and olivine in the calc-silicate ribbon produces the rusty staining. Warrensburg, Adirondacks, New York, USA.

Figure 5.11 Calc-silicate rock that occurs at the contact between diopside – quartz marble and metamorphosed gabbro, which are about a half meter away to the left and right, respectively. The marble and gabbro themselves have metamorphic mineral assemblages characteristic of the upper amphibolite facies. The calc-silicate rock shown here is composed of orange grossular-andradite garnet, dark-green diopside, and white scapolite and wollastonite. Where fresh, scapolite and wollastonite can be difficult to tell apart in the field. However, the greater solubility of wollastonite causes the initially rough surfaces to become smooth. The white material in most of the image is scapolite, but the smooth-looking white and gray material in the lower left corner is wollastonite. Strangely, the wollastonite-bearing assemblage shown here is characteristic of high-temperature, low-pressure contact metamorphism, not regional amphibolite facies conditions (Figs. 1.1, 5.1). At this outcrop common marble is abundant, and this anhydrous calc-silicate assemblage occurs only at the gabbro-marble contact. It is possible that the calc-silicate rock was originally formed when the gabbro intruded at low pressure, forming the dry calc-silicate contact metamorphic assemblage. This assemblage then survived subsequent upper amphibolite facies regional metamorphism. Comstock, Adirondacks, New York, USA.

Figure 5.12 This is an elliptical exposure of a calc-silicate rock (center) made mostly of quartz, diopside, and actinolite. The calc-silicate rock is surrounded by migmatitic quartz – feldspar – mica – sillimanite gneiss. On the outcrop, several possible explanations were offered for the shape of the calc-silicate body, including a large, pre-metamorphic concretion, or an unusually smooth boudin. As some at the outcrop suggested, it is most likely a tightly folded calc-silicate layer having a doubly-plunging fold hinge. That is, a folded-over layer having a nearly vertical axial surface, and a fold axis that is bent like a bow so that it plunges down into the gneiss at both ends. Milton, New Hampshire, USA.

Chapter 6

Mixed sedimentary rocks

Many sediments can be referred to as sandstones, shales, or limestones, but others are mixtures between those and other sedimentary end members. Siltstones, for example, generally have compositions between typical sandstone and shale, and marls have compositions between shale and limestone. Where metamorphosed, these rocks develop mineralogy and appearance that can be quite different from the metamorphosed end members. This has been discussed to some degree in Chapters 4 and 5 regarding marbles and some calc-silicate rocks. This chapter illustrates such sedimentary mixtures as they occur in metamorphosed stratigraphic sequences.

Differences in mineralogy can be expressed in terms of some example reactions. Lets take a simple example mixture of calcite and kaolinite clay. Calcite alone would metamorphose to calcite marble, and kaolinite alone would metamorphose to an unusual aluminosilicate-quartz rock. Mixtures, however, might undergo the following reaction:

$$\underset{\text{kaolinite}}{Al_2Si_2O_5(OH)_4} + \underset{\text{calcite}}{CaCO_3} = \underset{\text{anorthite}}{CaAl_2Si_2O_8} + \underset{\text{fluid.}}{CO_2 + 2H_2O} \quad (6.1)$$

Anorthite would not occur in either end member at any metamorphic grade, and so is unique to the mixture. Under different metamorphic conditions kaolinite + calcite might react to form a different mineral:

$$\underset{\text{kaolinite}}{3Al_2Si_2O_5(OH)_4} + \underset{\text{calcite}}{4CaCO_3} = \underset{\text{zoisite or clinozoisite}}{2Ca_2Al_3Si_3O_{12}(OH)} + \underset{\text{fluid.}}{4CO_2 + 5H_2O} \quad (6.2)$$

In this case, too, zoisite or clinozoisite are unique to the mixture and could not occur in either end member. Next, replace some of the calcite with dolomite and include some quartz:

$$\underset{\text{kaolinite}}{13Al_2Si_2O_5(OH)_4} + \underset{\text{dolomite}}{4CaMg(CO_3)_2} + \underset{\text{calcite}}{13CaCO_3} + \underset{\text{quartz}}{8SiO_2} =$$

$$\underset{\text{zoisite or clinozoisite}}{13Ca_2Al_3Si_3O_{12}(OH)} + \underset{\text{diopside}}{4CaMgSi_2O_6} + \underset{\text{fluid.}}{21CO_2 + 26H_2O} \quad (6.3)$$

Again, none of the solid reaction products could occur in a non-calcareous schist or a clay- and quartz-free dolomitic marble. Here is an example with illite as the clay mineral instead of kaolinite:

$$\underset{\text{illite}}{2K_{0.5}Al_2Si_{3.5}Al_{0.5}O_{10}(OH)_2} + \underset{\text{calcite}}{2CaCO_3} = \underset{\text{anorthite}}{2CaAl_2Si_2O_8} + \underset{\text{K-feldspar}}{KAlSi_3O_8} + \underset{\text{fluid.}}{2CO_2 + 2H_2O} \quad (6.4)$$

A low- or medium-grade, non-calcareous schist would be unlikely to have K-feldspar and could not have anorthite, and a clay-free marble could not have any feldspars. If the initial carbonate is dolomite:

$$\underset{\text{illite}}{2K_{0.5}Al_2Si_{3.5}Al_{0.5}O_{10}(OH)_2} + \underset{\text{dolomite}}{3CaMg(CO_3)_2} = \underset{\text{anorthite}}{2CaAl_2Si_2O_8} + \underset{\text{phlogopite}}{KMg_3Si_3AlO_{10}(OH)_2} + \underset{\text{calcite}}{CaCO_3} + \underset{\text{fluid.}}{5CO_2 + H_2O} \quad (6.5)$$

In this case the resulting rock could be mica-rich and schistose, but anorthite- and calcite-bearing, unlike either non-calcareous schist or clean dolomite marble. The mineralogy of metamorphosed carbonate-shale mixtures will therefore tend to have anorthitic feldspar and a variety of Ca-Al, Ca-Mg, and K-Mg minerals such as epidote, diopside, phlogopite, and Ca-amphibole, depending on the starting composition, metamorphic conditions, and fluid composition. You can play this reaction game using any end member rocks, with the prize being new mineral assemblages in the mixtures that would be impossible otherwise.

Figure 6.1 illustrates four end members common in sedimentary successions, and typical rocks at medium metamorphic grade that will form from the six binary

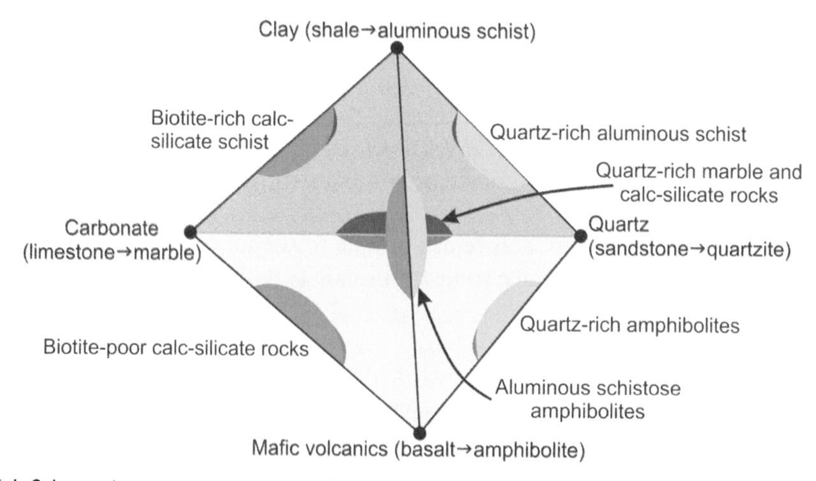

Figure 6.1 Schematic quaternary system showing four common sedimentary rock end member components that can end up in mixed sedimentary rocks. The end members (black dots) are labeled

Figure 6.1 (Continued) with the principal materials they represent. In parentheses the sedimentary rock protolith names are given, along with their medium-grade metamorphosed equivalents. The six fields along the edges (binary mixtures) are labeled with the medium-grade rocks these intermediate mixtures characteristically form. You can use your imagination to consider what kinds of rocks three- and four-component mixtures would turn into during metamorphism.

mixtures. Some of the reactions to make the mineral assemblages in intermediate rocks were given above, but readers are left to ponder the rest. Identifying which components, and how much of each, in rocks that had mixed sedimentary protoliths is basically a chemical mixing problem. It can be solved by defining the end member compositions, and then calculating how much of each is needed to make the rock composition you actually have. Rock compositions can be analyzed directly with a bulk chemical analysis, or calculated from analyzed or estimated mineral compositions and mineral modes. The point is that these are distinctive rocks, worthy of close examination.

In the illustrations below, emphasis is on distinctive appearance of mixed sedimentary rocks in the field, rather than metamorphic grade. The rocks shown in thin section (Fig. 6.2) were all metamorphosed to epidote amphibolite facies conditions, and so emphasize the mineralogical and textural differences between three different protolith mixtures. The field photos show rocks formed from a variety of mixture types that were metamorphosed in the epidote amphibolite and amphibolite facies. Mixtures include basalt-shale (Fig. 6.3), basalt-sand or silt-carbonate (Figs. 6.4, 6.8), basalt-shale-carbonate (Figs. 6.6, 6.9, 6.10), rhyolite-carbonate (Fig. 6.5) and more complex mixtures (Fig. 6.7). The different mixtures show a wide diversity of textures and mineralogy. In some cases it is the diversity of a geologic unit that is its distinguishing characteristic.

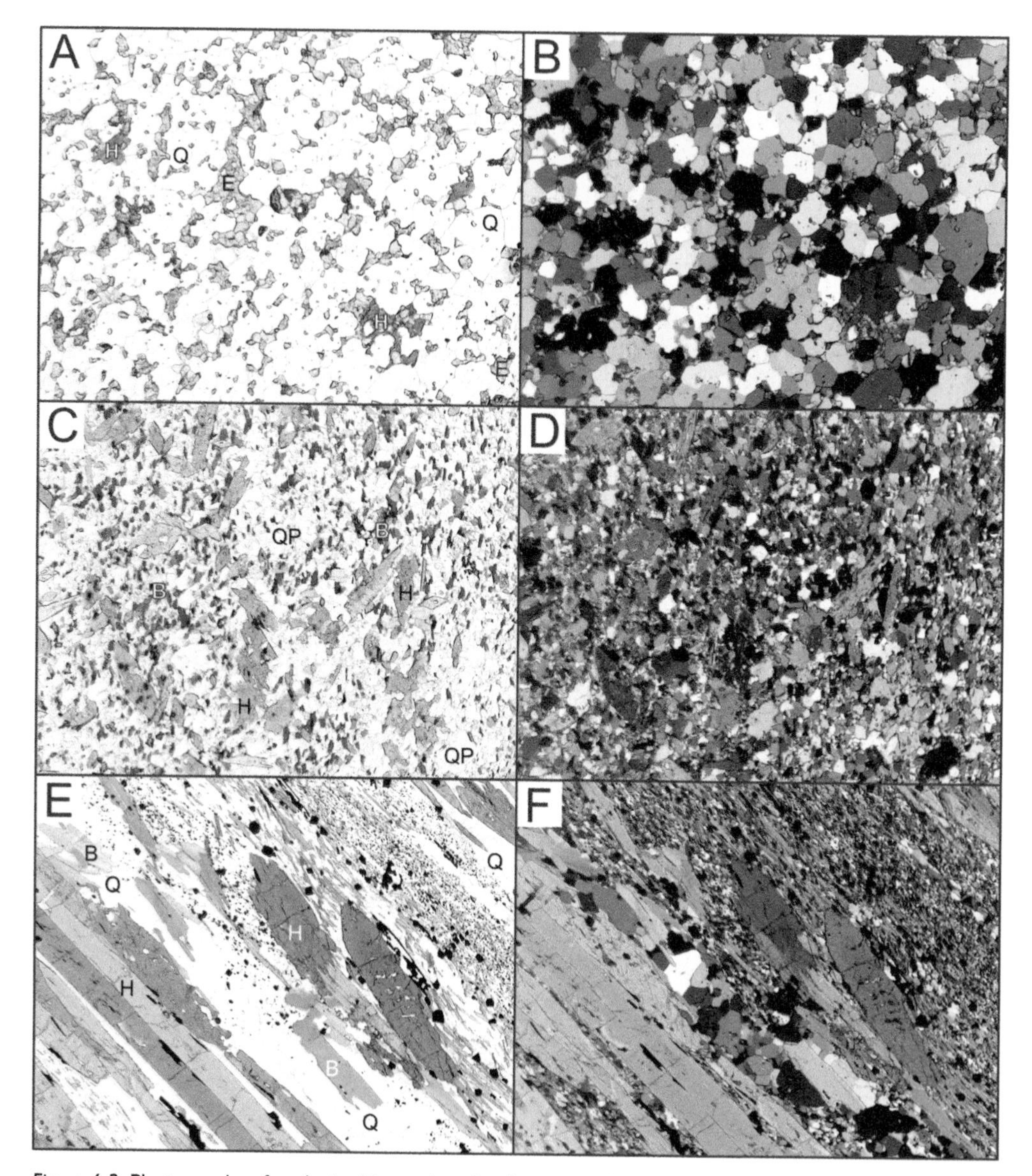

Figure 6.2 Photographs of rocks in thin section that have mixed sedimentary protoliths. All images on the left are in plane-polarized light, and those on the right are the same fields in cross-polarized light. Field widths are all 4 mm, and all rocks were metamorphosed at epidote amphibolite facies conditions. A, B) Hornblende – epidote – quartz rock, from a position in Figure 6.1 close to the clay-carbonate-quartz face of the quaternary system. The lack of mica suggests that the clay component was kaolinite. Lebanon, New Hampshire, USA. C, D) Hornblende – biotite – quartz – calcic plagioclase rock, from a position in Figure 6.1 close to the clay-carbonate-volcanics face. Candia, New Hampshire, USA. E, F) Hornblende – biotite – quartz rock, from a position in Figure 6.1 on the clay-quartz-volcanics face, probably much closer to the volcanics-quartz line than to clay. Charlemont, Massachusetts, USA. Abbreviations: B, biotite; E, epidote; H, hornblende; Q, quartz; QP, quartz and plagioclase.

Figure 6.3 Hornblende bundles in a quartz – biotite – muscovite – garnet schist. The hornblende crystals nucleated at single points, and grew along their c-axes mostly in the plane of the rock foliation and cleavage. This kind of texture is commonly referred to as garbenshiefer, from the German for sheaf-schist (as in a sheaf of wheat). Notice the two garnets near the upper right corner (red arrows). Charlemont, Massachusetts, USA.

Figure 6.4 Quartz – garnet – hornblende rock that was probably a mixture between basaltic volcaniclastics and small amounts of other sedimentary material that probably included quartz and carbonate. The weathered out pits to the left (yellow arrow points to one) were once occupied by carbonates. Notice how the large, brown garnets have overgrown the rock foliation (one example garnet indicated with a red arrow, with a red line parallel to the foliation). Brockways Mills, Vermont, USA.

Figure 6.5 Calcareous schist (brownish layers, indicated by red arrows) between layers of rhyolite (probably volcaniclastic), metamorphosed to epidote amphibolite facies. The schist contains 10-50% brown-weathering carbonate with quartz and white mica. Considering the adjacent layers, this rock probably had a protolith that was a carbonate-rhyolite volcaniclastic mixture. The white areas are deformed and boudinaged quartz veins. Brattvåg, western Moldefjord, Møre og Romsdal, Norway.

Figure 6.6 Amphibolite (red arrow), and interlayered brownish biotite – hornblende – plagioclase and gray-green diopside – quartz – plagioclase – biotite layers (blue arrow). A thick diopside-rich layer (yellow arrow) in the interlayered sequence is boudinaged, indicating that it was more brittle than the surrounding thinner layers. White arrows point to boudin necks, which have white vein quartz in them. These rocks were probably derived from metamorphosed sediments made from variable proportions of clay, carbonate, and basaltic volcaniclastic components. The white layers are deformed and boudinaged quartz veins. Midøy, Midsund, Møre og Romsdal, Norway.

Figure 6.7 Interlayered epidote amphibolite (top only, A), hornblende-rich layers (gray, H), diopside-rich layers (greenish-gray, D), and mica schist (rusty staining, S). Thicker layers are labeled. The layers, and possibly even the amphibolite, are probably metamorphosed sediments that contained variable quantities of carbonate, clay, quartz, and volcaniclastic components. The rusty stain is caused by weathering sulfides in the schist layers, which suggests that the clay-rich component was deposited under reducing conditions. Though the sedimentary sequence shown here is only about 10 cm thick, it may once have been many meters thick prior to deformation. Kvithylla, western Trondheimsfjord, Sør Trøndelag, Norway.

Figure 6.8 Layered mixed sediment that has been metamorphosed to epidote amphibolite facies conditions. The layer labeled 1 is coticule, pink garnet quartzite, in which the typical garnet size is only 1-5 μm. Coticule is generally interpreted to be metamorphosed Fe- or Mn-bearing chert. The layers labeled 2 and 4 are made of quartz, hornblende, biotite, ankerite, and epidote, and are probably a carbonate-volcaniclastic-quartz sand (or chert) mixture. Many of the ankerite crystals are euhedral and rhomb-shaped, and weather out to form rhomb-shaped pits. The layer labeled 3 is a feldspathic quartzite having the assemblage quartz – plagioclase – biotite – ankerite. The white layer on the far right is a quartz vein. Charlemont, Massachusetts, USA.

Figure 6.9 This is a complex section composed of amphibolites (upper right), inhomogeneous, brown biotite- and hornblende-rich rocks (mostly to the lower left), and the green epidote – diopside – quartz – carbonate-rich rock layers passing diagonally across the image (possibly one folded layer). Although the green layers may be a folded hydrothermal vein, in this somewhat calcareous, volcanic-rich, layered sequence, they are more likely to have been a sedimentary mixture between carbonate and volcaniclastic material. If so, the volcaniclastic material must have lost its sodium and potassium, possibly during weathering, which may have occurred in an oxidizing submarine environment. The outcrop, particularly on the left side, has been cut by a series of late, crosscutting fractures that are filled with thin quartz veins. Årnes, Surnadalsfjord, Møre og Romsdal, Norway.

Figure 6.10 Interlayered dark-gray amphibolite, thin brownish biotite-rich layers, and greenish epidote – diopside – quartz layers. The whole section probably represents metamorphosed mixtures between volcaniclastic, carbonate, quartz sand, and shale sedimentary components. The flat green patch, outlined by a white dotted line, is a shiny slickensided surface that is covered with epidote and chlorite. Midøy, Midsund, Møre og Romsdal, Norway.

Chapter 7

Conglomerates and breccias

Conglomerates generally indicate rapid deposition from nearby elevated terranes, such as a mountain range, a rift valley flank, or beach escarpment. They are therefore useful for interpreting local paleogeography, and more indirectly the tectonic processes responsible for it. Polymictic conglomerates have a variety of clast rock types (e.g., a mixture of volcanic rocks, plutonic rocks, and sandstone), while those consisting of only a single type of clast (e.g., only granite or only vein quartz) are monomictic. Clast types tell about the eroding, uplifted source terrain. For example volcanic arcs are dominated by volcanic and low-grade metamorphic rocks, deeply eroded continents are dominated by plutonic and medium- to high-grade metamorphic rocks, and regions draining carbonate rock escarpments will tend to be dominated by limestone and chert clasts.

Strongly anisotropic or soft rocks like phyllite and schist tend to fragment easily, and so are uncommon in conglomerates except very close to the source region. The quick destruction of well-foliated rocks prior to conglomerate deposition means that conglomerate clasts tend to be more or less equant in shape, rather than platy or pencil-shaped. Because of their roughly equant initial shapes and generally minor depositional preferred orientation, deformed clasts can be useful strain markers for structural geologic studies. During deformation, roughly spherical clasts (at least on average) take on shapes that can be approximated by triaxial ellipsoids, in which the lengths of the three mutually perpendicular axes represent the deformed clast shape. Clasts can be extended in one direction to produce a lineation, flattened to produce a foliation, or extended and flattened to produce a rock with both.

Breccias begin with rock fracturing into angular fragments, commonly as a result of brittle behavior caused by low temperatures or high strain rates. The matrix to the angular fragments can be another variety of rock, into which the breccia fragments are incorporated, or it can be a more finely broken form of the same rock.

In some cases deformed clasts may be difficult to tell from boudins (Chapter 18), particularly if the objects are sparse, or if the boudins become anomalously rounded. Figure 7.1 shows some examples of the transformation of clasts into boudins (Figs. 7.1A, C), and layers into similar-looking boudins (Figs. 7.1B, D). In most cases, paying close attention to rock type, boudin spacing, and boudin tails should make it possible to distinguish between clasts and boudins. Deformation can conspire to make distinguishing the two more difficult, however, in isolated cases (Fig. 7.1E, F).

In this section there are examples of both metamorphosed conglomerates and breccias. The conglomerates, Figures 7.2–7.10, are shown in a sequence of increasing metamorphic grade: greenschist facies, epidote amphibolite facies, and amphibolite facies.

Figures 7.10–7.12 show a strongly deformed quartz pebble conglomerate on three mutually perpendicular surfaces. Figures 7.13–7.15 show metamorphosed igneous and tectonic breccias at lower and upper amphibolite facies.

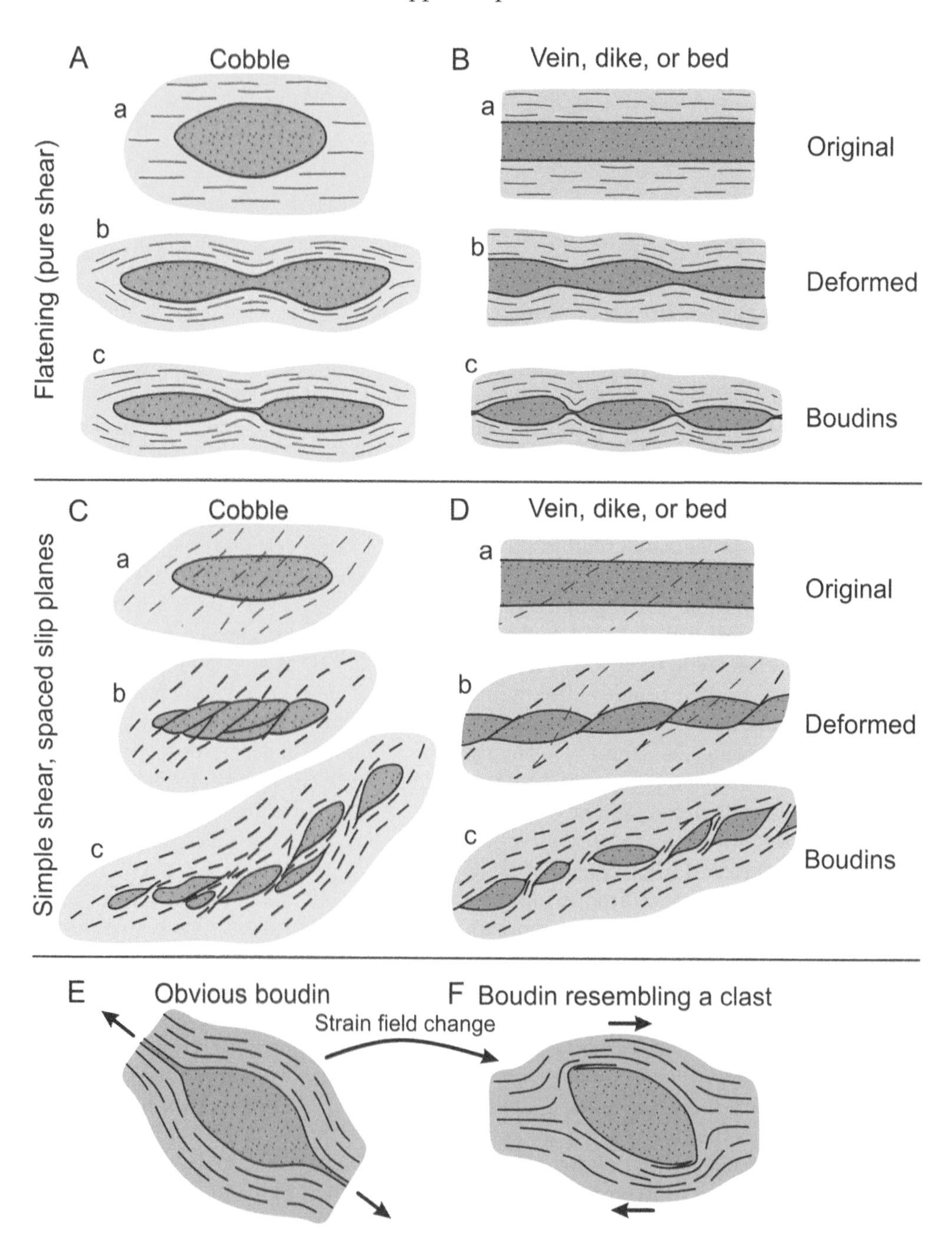

Figure 7.1 Conglomeratic clasts can, in some cases, be difficult to distinguish from boudins, as illustrated in these schematic examples. A and B represent a conglomeratic clast and a layer,

Figure 7.1 (Continued) respectively. Progressive deformation of each can produce an odd-looking, partially boudinaged clast or bed, and strings of boudins. In C and D, spaced cleavage surfaces (effectively local shear zones) disrupt and offset clast and bed fragments to produce boudins that don't have the characteristic blocky ends or connecting tails, making them more difficult to tell from clasts. E and F show an example of how tails on ductile boudins (E) can themselves be deformed and attenuated so that they lose their obvious identity (F, see Fig. 18.11). The point is that boudins can be confused with clasts, so how does one tell a deformed conglomeratic clast from the odd boudinaged lump? Most boudins develop from competent layers, dikes, or veins, and so tend to form strings of lithologically identical pieces, commonly either immediately adjacent to or connected by thinned necks. Clasts, however, are usually deposited with a great many peers, and so yield a haphazard arrangement with no blocky ends or connecting tails. Clasts may be deformed, but will usually be arranged with other, similar-looking objects packed together. If boudinaged, adjacent pieces will be the same rock type, but will not form long strings of regularly spaced boudins. Not every rounded lump is a clast, but the presence of boudins should not negate the possibility of a conglomeratic protolith. It is wise to use context, for example tracing a unit that varies along strike from obvious, little deformed conglomerate to more severely deformed rocks. A-D are modified after Spry and Burns (1967).

Figure 7.2 This is a glacially polished outcrop surface showing conglomerate that has undergone essentially no deformation, though it is at greenschist facies. This conglomerate contains a wide variety of sub-rounded clasts including felsic and mafic volcanics, chert, and quartzite. Notice that there is little or no preferred orientation of the clasts, typical of undeformed conglomerates. Caucomgomoc Lake area, Maine, USA.

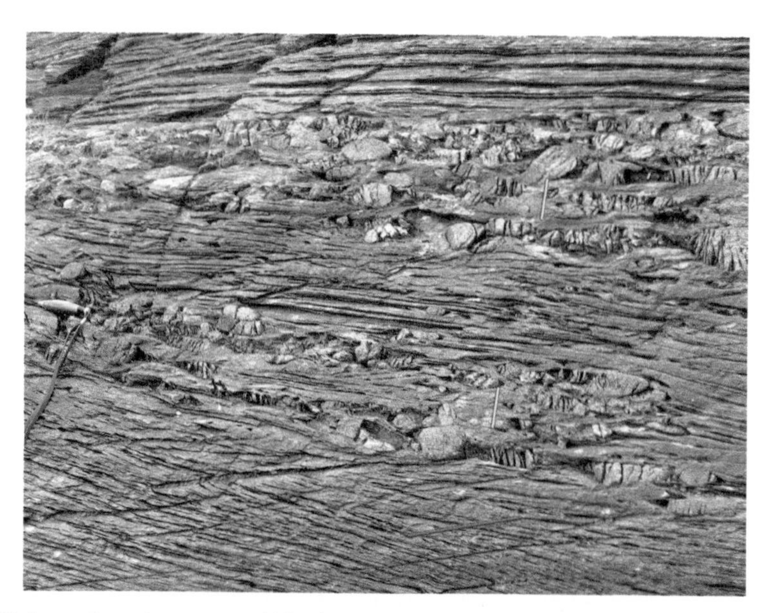

Figure 7.3 Deformed conglomerate cobbles (granite, vein quartz, volcanic rocks) in a muddy sandstone matrix, metamorphosed to middle greenschist facies. This rock is interpreted to be an Ordovician turbidite sequence that was deposited offshore from a nearby highland. The two conglomerate layers visible here are actually one, repeated across the axial surface of a tight fold with the hinge to the left. The prominent set of lines in the fine-grained rock (approximately parallel to the blue pen) is the axial planar cleavage to the isoclinal folds, and so it is approximately parallel to the folded layering. Extension of the clasts parallel to the cleavage resulted in sets of extensional fractures in many of the clasts, parallel to the yellow lines. There is also a sparse spaced cleavage that extends from the upper right to lower left, parallel to the red line, that crosses the conglomerate layers and earlier cleavage. Storvika, Stjørdal, Nord Trøndelag, Norway.

Figure 7.4 Deformed cobble conglomerate, showing large, distorted quartzite cobbles in a matrix of more severely deformed sand and gravel-size clasts. This rock has been metamorphosed to greenschist facies conditions. Goat Island, Tasmania.

Figure 7.5 Muddy sand-matrix conglomerate, metamorphosed at greenschist facies. The clasts weather out in raised relief, clearly showing their outlines. Some of the clasts contain fractures that are perpendicular to the extension direction (red arrows show examples of clasts with such fractures, yellow arrow shows the extension direction). Just above the wooden bench are quartz veins (blue arrows) that opened in the extension direction, like fractures in the clasts. Blekkpynten, Fættenfjord, Nord Trøndelag, Norway.

Figure 7.6 Deformed conglomerate, containing mostly quartzite clasts but with a wide variety of minor rock types, metamorphosed to upper greenschist facies. The clasts have been extended in a direction parallel to the red arrow. Although the individual conglomerate clasts weather out separately, possibly suggesting that this unit has been little-deformed, it is actually part of an overturned fold. The field width is about 2.5 m. Newport, Rhode Island, USA.

Figure 7.7 Deformed conglomerate metamorphosed to epidote amphibolite facies conditions. The clasts include white and gray quartzite, which is easy to see, and granitic rocks which tend to look much like the fine-grained matrix (two pointed out with red arrows). All have been extended in approximately the vertical direction. Florida, Massachusetts, USA.

Figure 7.8 This is a quartz pebble conglomerate from the same unit (Silurian Clough Quartzite) as Figures 7.9–7.12. Here it has been metamorphosed to greenschist facies conditions and, though involved in folding, appears not to be highly deformed at this location. Notice the roughly even distribution of clast sizes from large to small. Littleton, New Hampshire, USA.

Figure 7.9 Quartzite from the same unit as Figure 7.8, but metamorphosed to lower amphibolite facies conditions. In this case the rock has rather sparse large clasts (ovoid white spots), in a finer-grained, somewhat graphitic quartz-rich matrix. The rock looks little-deformed, but in fact it has a weakly-developed foliation parallel to the red lines. The large, irregular white masses at the top are early, deformed quartz veins, and the white line extending from lower left to upper right is a late, relatively undeformed quartz vein. Black Mountain, New Hampshire, USA.

Figure 7.10 Quartzite from the same unit as Figures 7.8 and 7.9, metamorphosed to amphibolite facies conditions and strongly deformed. This is the first of three images (7.10–7.12) taken of the same outcrop

Figure 7.10 (Continued) on three mutually perpendicular surfaces This surface is parallel to the foliation and lineation. The foliation is defined by the flattened pebbles and rare muscovite, and the lineation is parallel to the pen. Individual pebbles, such as the white one to the right of the pen, have been deformed so that original clasts, once approximately spherical, now have ellipsoidal axis proportions of approximately 50:5:1. The inset shows a schematic of this surface and pen location. Northfield, Massachusetts, USA

Figure 7.11 Second of three images (7.10–7.12) of quartz pebble conglomerate, this one showing a surface broken perpendicular to foliation and parallel to lineation, both of which are parallel to the pen. The inset shows a schematic of this surface and pen location. Northfield, Massachusetts, USA

Figure 7.12 Third of three images (7.10–7.12) of quartz pebble conglomerate, showing a surface perpendicular to foliation and lineation. The pen is in the foliation plane and perpendicular to the lineation. The thick, white layer at the top is a quartz vein, not a deformed clast. The inset shows a schematic of this surface and pen location. Northfield, Massachusetts, USA.

Figure 7.13 This is a deformed xenolith-rich intrusion breccia, looking parallel to the lineation. All of the clasts are dark-gray hornblende diorite in a light-gray tonalite matrix. Although the original texture is igneous, these rocks were deformed and metamorphosed in the epidote amphibolite facies. In this view, xenoliths partly retain their original angular shapes and inconsistent orientations, though they have been extended on this surface parallel to the red arrow. Different perspectives with respect to deformational fabrics are important for interpreting them. Almvikneset, outer Trondheimsfjord, Sør Trøndelag, Norway.

Figure 7.14 Tectonic breccia at upper amphibolite facies. This consists of fragments of light-colored tonalite and pegmatite in a highly strained, dark-gray garnet – biotite tonalite matrix. Crosscutting and other relationships indicate that deformation disrupted multiple generations of tonalite and pegmatite dikes and sills. Several late dikes (some highlighted with red arrows), dipping to the right, are relatively intact and so post-date most of the deformation (see Morton, 1985). Phillipston, Massachusetts, USA.

Figure 7.15 Tectonic breccia of amphibolite, vein quartz, and calc-silicate fragments in marble metamorphosed in the upper amphibolite facies. As mentioned in Chapter 4, marble is very ductile at high metamorphic grades, and so can take up a lot of strain compared to surrounding, stronger, silicate rocks. Fragmentation of layers in the marble, and rocks at its contacts, can choke the marble with angular fragments of a wide variety of rock types. Such marbles look much like igneous breccias, but they formed entirely in the solid state. Paradox Lake, Adirondacks, New York, USA.

Chapter 8

Gneisses

Gneisses are generally poorly foliated, medium- to coarse-grained feldspathic rocks. As such, they plot close to the feldspar corner of mineral proportion space: too much amphibole and they tend to be called amphibolite, too much quartz and they become quartzite, and too much mica and they become schist, for example. The poor foliation of gneiss is typically a consequence of having only small proportions of sheet silicates like muscovite and biotite, rather than from lack of strong deformation.

Gneisses are commonly derived from felsic or intermediate plutonic rocks or their volcanic equivalents, but they can also be derived from particularly feldspathic arkoses, or even shales if the metamorphic grade is high enough to dehydrate most sheet silicates. Gneisses may be the most abundant rock type in continental crust, and vast areas of ancient continental shields are certainly dominated by gneissic rock. This stems from the fact that, in most continental areas, sedimentary and mafic volcanic rocks form a comparatively thin veneer on continental surfaces. Even where sediments are thick, such as on continental margins, much of that material is thinned during plate tectonic collisions and metamorphism, and much of that material is uplifted and eroded away to sediment. The deeper crust is dominated by plutonic and medium- to high-grade metamorphic rocks. The felsic and intermediate plutonic rocks, once deformation begins, become gneisses too, and thus a subject of this chapter.

Gneiss protoliths may have been highly homogeneous at large scales, such as large bodies of granite or diorite, or they may have may have had inhomogeneities such as sedimentary bedding, crosscutting dikes, or xenoliths. Inhomogeneities can also develop during metamorphism, such as by segregation of partial melts, or metamorphic differentiation during folding and cleavage development. With deformation, all these inhomogeneities tend to become folded layers, so their origin can be difficult to interpret. Figure 8.1 shows some examples of a homogeneous igneous rock (gray) containing felsic igneous inhomogeneities: pegmatite dikes (A, B), xenoliths (C, D), and melt segregations produced during metamorphism (E, F). After severe deformation, the original felsic rocks form a series of tightly folded layers that are not easy to distinguish in terms of their original nature. One might have to look carefully at chemical compositions, or in less deformed areas such as fold hinge regions, to determine the structural relationships.

Figure 8.2 shows images of gneisses in thin section. The first three image pairs show granitic gneisses in order of increasing metamorphic grade from epidote amphibolite facies to transitional between amphibolite and granulite facies. The last pair shows a gneissic aluminous rock metamorphosed to high metamorphic grade, lacking the muscovite and some biotite that it would have at lower grade. The first two field photos (Figs. 8.3, 8.4) show the effects of different amounts of deformation on what were once probably almost identical granites. The next set (Figs. 8.5–8.7) shows gneisses that have experienced relatively low-grade metamorphism and so have not experienced any partial melting. The next two are transitional, having melted to only a small degree (about 1–2%, Figs. 8.8, 8.9). The third set have experienced large amounts of partial melting (about 5–20%, Figs. 8.10–8.15). Partial melting produces silicate liquid, essentially magma, which can flow into fractures or other opening spaces. This separation of melt from residual solid rock is called melt segregation. The result is a mixed rock with a metamorphic part (restite, the residual solid) and the igneous part (the liquid, now crystallized to rock). Such rocks are called migmatites and are important enough in high-grade terranes that they have their own chapter (21). Many migmatites are also gneisses, so here is your first introduction to them.

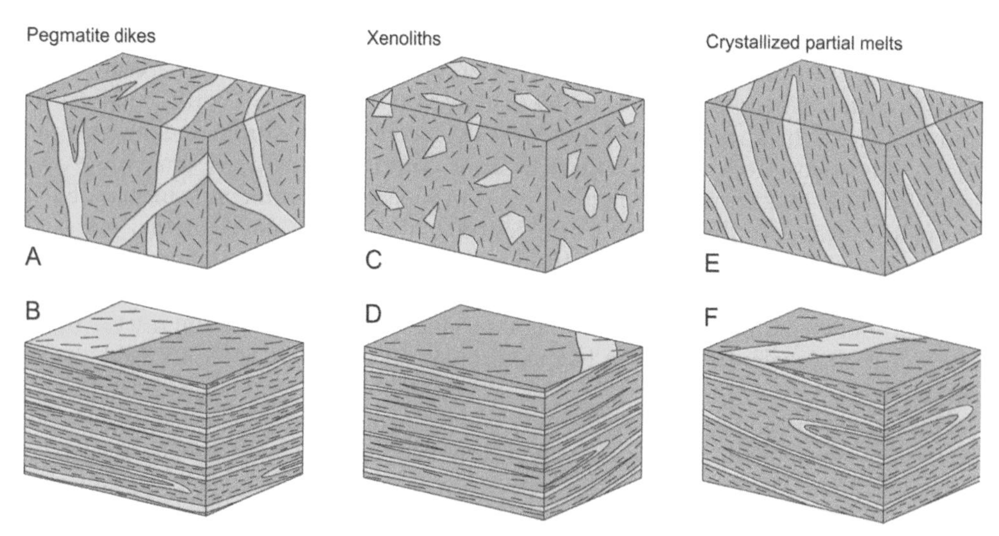

Figure 8.1 Illustration of some ways gneissic rocks can get their layering. The top row shows initial rocks, and the bottom row shows the resulting appearance after deformation. A and C are entirely igneous, with a homogeneous rock (say, diorite, gray) hosting pegmatite dikes (A) or granite xenoliths (C). E is already a gneissic metamorphic rock (gray, say, a diorite gneiss) that has partially melted, with the granitic liquid having segregated into discontinuous layers where it crystallized. After strong deformation, it is difficult to tell what the granitic layers in the three different gneisses, B, D, and F, originally were prior to deformation.

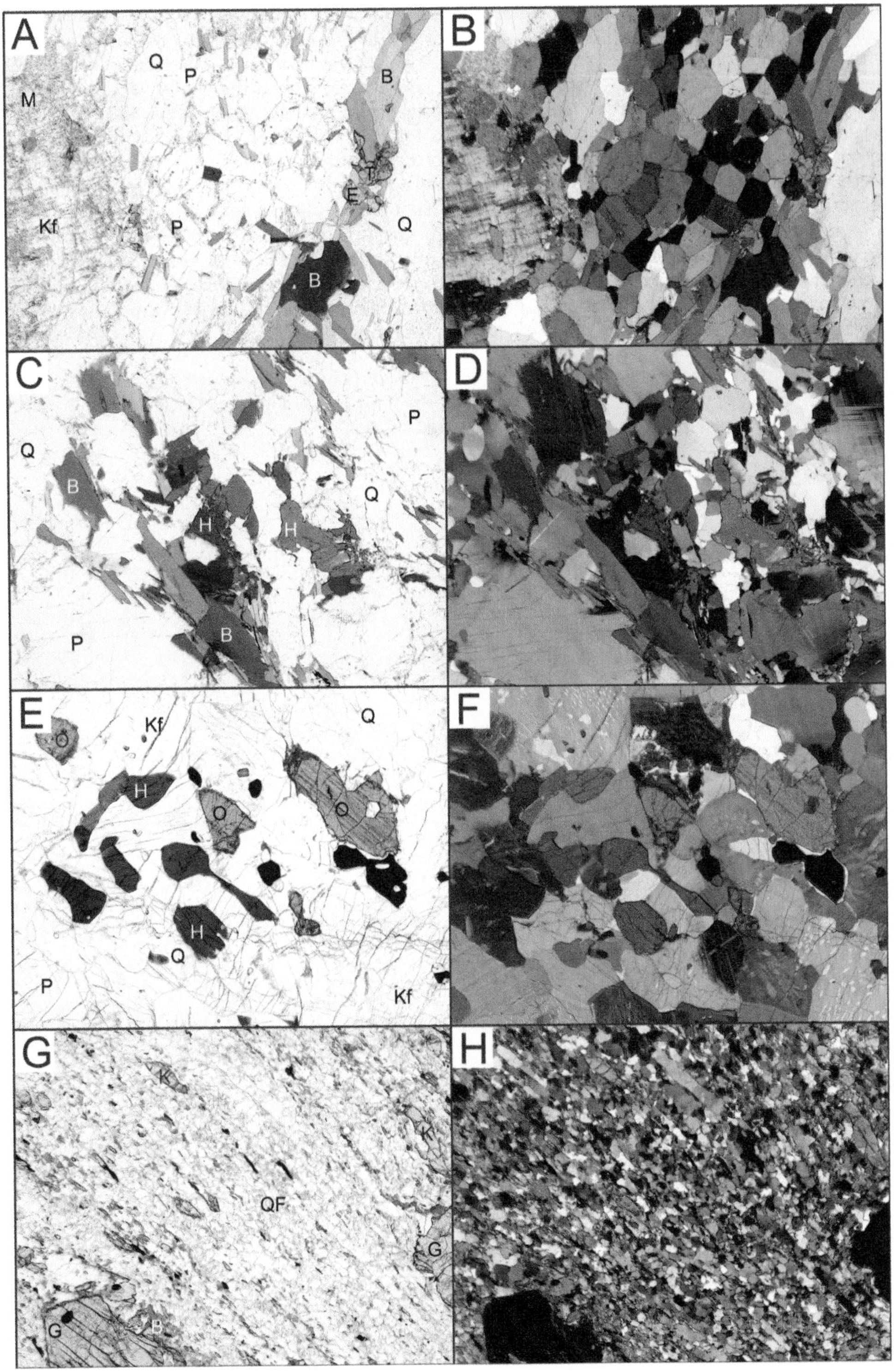

Figure 8.2 Photomicrographs of gneissic rocks in thin section. Images on the left are in plane-polarized light, and those on the right are the same fields in cross-polarized light. All field

Figure 8.2 (Continued) widths are 4 mm. A, B) Poorly foliated granitic gneiss, with the epidote amphibolite facies assemblage quartz – plagioclase – microcline – biotite – epidote – titanite, plus a little retrograde muscovite. Pike, New Hampshire, USA. C, D) Poorly foliated tonalitic gneiss, with the amphibolite facies assemblage quartz – plagioclase – biotite – hornblende. Quabbin Reservoir, Massachusetts, USA. E, F) Essentially unfoliated charnockitic gneiss (orthopyroxene-bearing granitic gneiss) with the transitional granulite facies assemblage quartz – microperthite – plagioclase – hornblende – orthopyroxene. In outcrop this rock has strongly deformed and folded layers, but recrystallization and grain growth after deformation has erased most foliation at thin section scale. Notice the exsolved albite (F, light-gray, irregular spots) in darker-gray K-feldspar. Schroon Lake, Adirondacks, New York, USA. G, H) Moderately foliated aluminous gneiss, probably metamorphosed shale. It has the assemblage quartz – microcline – plagioclase – garnet – kyanite – biotite. Despite its origins as a shale, the scarcity of micas makes this rock gneissic rather than schistose. Fjørtoft, Nordøyane, Møre og Romsdal, Norway. Abbreviations: B, biotite; E, epidote; G, garnet; H, hornblende; K, kyanite; Kf, K-feldspar; M, muscovite; O, orthopyroxene; P, plagioclase; Q, quartz; QF, quartz and feldspar; T, titanite.

Figure 8.3 Coarse-grained, relatively undeformed granitic gneiss, metamorphosed to epidote amphibolite facies. Large pink K-feldspar porphyroclasts are visible (former igneous phenocrysts) in a finer-grained matrix, representing a relict igneous texture. The matrix has white plagioclase, bluish quartz, K-feldspar, and a black mixture of biotite and epidote. Uthaug, Sør Trøndelag, Norway.

Figure 8.4 A more severely deformed gneiss that was originally probably like that in Figure 8.3. The relict igneous phenocrysts (porphyroclasts in 8.3) are now augen: eye-shaped K-feldspar porphyroclasts that have been recrystallized on their margins and smeared out into the foliation plane (see Chapter 17). Two augen are indicated by red arrows, but many more of pink K-feldspar and white plagioclase are visible. Hejnskjel Island, Hagaskjera, Sør Trøndelag, Norway.

Figure 8.5 Strongly deformed, layered tonalitic gneiss. The layering represents smeared out inhomogeneities that include dikes and xenoliths, which are recognizable in less deformed parts of this same outcrop. This is an example of deformational effects seen in Figures 8.1A-B and C-D. Baksteinen Peninsula, Sør Trøndelag, Norway.

Figure 8.6 Strongly deformed, layered tonalitic gneisses, with thick white and gray layers. The white layers are poor in biotite and have very little internal layering. The thick gray layers are themselves made up of alternating, thin medium-gray and white layers. Many of the thin white layers are made entirely of quartz, and so are deformed quartz veins. Other white layers seem to be tonalitic, like the thick white layers, and so may be interlayered volcanics or strongly deformed dikes. Some of the thin tonalitic white layers are slightly crosscutting (one example indicated by a red arrow) and so are apparently deformed dikes (an example of the deformational effects illustrated in Fig. 8.1A-B). From this outcrop the geologic sequence seems to have been: 1) formation of the gray rock, 2) emplacement of the quartz veins, 3) emplacement of the white tonalite as dikes, and 4) strong deformation. Råkvåg, Sør Trøndelag, Norway.

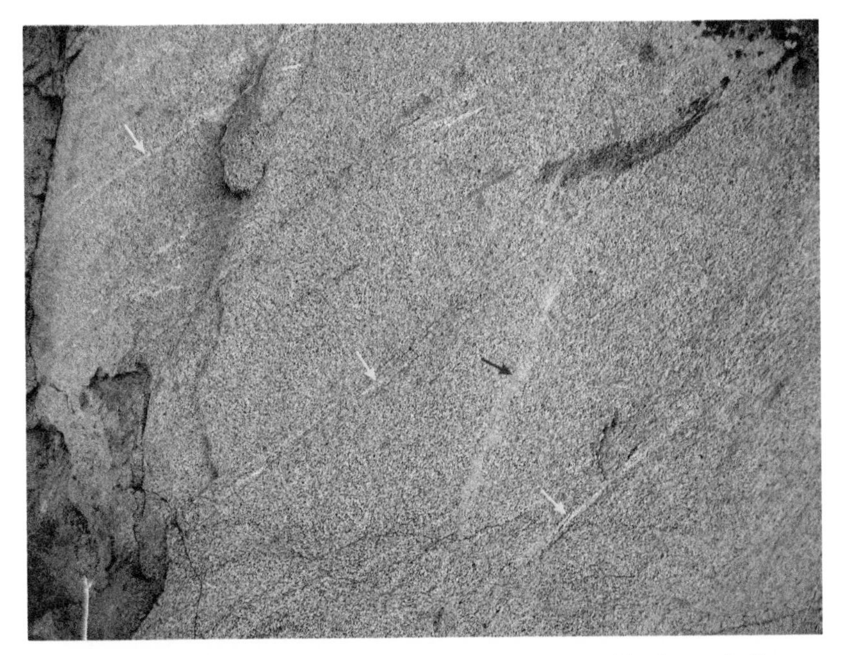

Figure 8.7 Coarse-grained tonalitic gneiss that has thin, discontinuous felsic layers (yellow arrows) and several small mafic enclaves, the largest of which is pointed out with a red arrow. The thin felsic layers may be deformed pegmatite dikelets, and the mafic enclaves may be deformed xenoliths. The cross-cutting pinkish body (black arrow) is fine-grained and granular, and looks on close examination like a deformed aplite dike. This is an example of the deformational effects seen in Figures 8.1A-B and C-D, but deformation was not severe enough to turn the enclaves into outcrop-spanning thin layers, or to make the pink dike nearly parallel to the other elongate features. Kjørsvik, Møre og Romsdal, Norway.

Figure 8.8 Charnockitic gneiss, cut by a basaltic dike (outlined in red). Charnockites are granites that contain orthopyroxene in addition to other mafic minerals such as hornblende and biotite. The fresh,

Figure 8.8 (Continued) unweathered, broken surfaces are gray. Orthopyroxene on exposed surfaces weathers quickly, releasing Fe^{2+}. The iron is oxidized and precipitated as the brown limonite staining seen on many of the flat joint surfaces. The gray to brown coloration of charnockitic gneiss outcrops helps differentiate them from other granitoid gneisses that are more typically white, light-gray, or pink in this part of the world. This rock has undergone partial melting, but the amount of melting was small and the melt segregations are almost the same color as the host rock, and so hard to see at this scale. Whitehall, eastern Adirondacks, New York, USA.

Figure 8.9 Granitic gneiss with small augen of pink K-feldspar and white plagioclase. There is a fine-grained layer under the toe of the boot, possibly a deformed aplite dike, and three light-colored, possibly pegmatitic layers or segregated local partial melts. All have been deformed so that the layers are parallel to themselves and to the rock foliation, though they are unlikely to have been parallel originally. Vingan, Sør Trøndelag, Norway.

Figure 8.10 Migmatitic granitic gneiss, with somewhat diffuse, patchy leucosomes (light-colored material, crystallized partial melt) that suggest that this rock was not strongly deformed after it was melted.

Figure 8.10 (Continued) Deformation tends to turn inhomogeneities like these into parallel layers, so this outcrop is intermediate between the undeformed and deformed melt segregations shown in Figures 8.1E and F, respectively. Warrensburg, Adirondack Mountains, New York, USA.

Figure 8.11 Complexly deformed gneiss containing boudins of eclogite (red arrows), amphibolite (black arrow), and lots of white material that is mostly highly deformed, crystallized segregations of partial melt, akin to Figure 8.1F. Ræstad, Otrøy, Møre og Romsdal, Norway.

Figure 8.12 Gray granitic gneiss with vertical layering that contains many black amphibolite boudins. Some of the amphibolites are folded (red arrows) with steep axial surfaces approximately parallel to the layering. The gneiss and amphibolites were intruded by granitic liquid, forming the white rock (yellow arrows) along sub-vertical channelways. Intrusion of the white rock disrupted some of the amphibolites even more than they had been previously. Putnam, eastern Adirondacks, New York, USA.

Figure 8.13 Migmatitic, garnet-rich granitic gneiss. The white partial melt segregations have clearly been folded (numerous fold hinges visible, two highlighted with red lines), indicating that melting preceded at least the last episode of folding. This is an example of the deformational effects illustrated in Fig. 8.1E-F. Juvika, Gossa Island, Møre og Romsdal, Norway.

Figure 8.14 Aluminous gneiss containing abundant garnet and small amounts of sillimanite and biotite, but no muscovite. This rock may have been derived from a compositionally layered shale or feldspathic siltstone, but here, metamorphosed to upper amphibolite facies conditions, all of the muscovite has dehydrated. The rock is migmatitic, having partially melted, and the numerous thin white layers are strongly deformed partial melt segregations like those shown in Figures 8.1E and F. About two-thirds of the way up the rock face the garnet-rich gneiss is cut by a white and pink granitic dike (red arrow), which has been deformed to be almost parallel to layering. Ticonderoga, eastern Adirondacks, New York, USA.

Figure 8.15 Complexly folded, migmatitic granitic gneiss. Visible are white and pink layers, some of which are apparently partial melts derived from the gneiss, and some are younger, crosscutting pegmatites presumably from outside of this outcrop (the largest, crosscutting pegmatitic body is indicated by a red arrow). There was likely more than one generation of partial melting that affected this outcrop. To the upper right are two amphibolite boudins (A). Unraveling the lithologic age and deformational and metamorphic history of outcrops like this takes careful observation in the field, thoughtful selection of samples, and a lot of lab work. It is from such evidence however, that the history of mountain belts is divined. Dryna, Midsund, Møre of Romsdal, Norway.

Chapter 9

Basaltic rocks, low and intermediate pressure

"...basaltic rocks are the solace of magmatists in these days of virulent transformism! One may derive comfort from the reflection that here at least are rocks which, to begin with, are of undisputed eruptive origin, of well-known and generally remarkably uniform chemical and mineralogical composition, and of familiar texture. Such comfort disappears soon enough after a contemplation of their metamorphic equivalents, here collectively called metabasaltic rocks." Poldervaart (1953)

Mafic rocks start out as lavas, dikes, gabbroic plutons, or pyroclastic deposits. Even where they have not been metamorphosed, the field term 'basalt' or 'basaltic' can, in many regions, include a rather broad range of compositions from picritic to andesitic, because it is hard to tell one fine-grained, dark rock from another. Prograde metamorphism can produce larger grain sizes, and results in mineralogical changes that, in many cases, gives some guidance to the chemical composition that a particular basaltic rock has.

For example, in a region where the rocks are at similar, medium metamorphic grades, the presence of the Mg-Fe amphiboles gedrite, anthophyllite, or cummingtonite imply more Mg-rich compositions, whereas garnet implies more Fe-rich compositions. The presence of Mg-Fe amphiboles along with an Al-rich mineral, like staurolite, cordierite, or Al-silicates, implies Ca-loss from the original rock by some alteration process (e.g., hydrothermal alteration, Fig. 1.9D). The presence of hornblende with more biotite than usual for the area, and visible quartz, may indicate an andesitic composition rather than basaltic. Diopside-bearing amphibolites may indicate a protolith enriched in cumulus clinopyroxene, or post-solidification addition of calcite. In the field you see rocks, visible minerals, and stratigraphy. In the lab you learn which rocks have particular mineral proportions, and what the chemical compositions are. Back in the field again, understanding from lab work can help aid rock identification and guide sampling programs.

More important for basaltic rocks than for many other kinds, the textural changes that occur during metamorphism can dramatically depend on the original hydration state of the rock. Granted, basaltic rocks are not generally thought of as being particularly hydrous, but they can certainly become so through a variety of processes including interaction with cold groundwater, hydrothermal alteration, and fluid infiltration from dehydrating rocks nearby. If the rock is hydrated prior to or early in the metamorphic history, the metamorphic mineral assemblage will include hydrous minerals like chlorite and white mica. These minerals are easily deformed, and so the rock itself will be easily deformed as well. In contrast, if the rock remains anhydrous, its original mineral assemblage (e.g., pyroxene – plagioclase – olivine – magnetite) is relatively strong, and during metamorphism may be transformed into other strong minerals (e.g., garnet). The result is that anhydrous basaltic rocks may be strong enough to retain their shape, and so retain some pre-metamorphic textures, while deformation takes place in more ductile surrounding rocks. Of course, some metamorphosed regions simply never experienced much deformation.

Hydrated basaltic rocks at low grade, containing chlorite and other sheet silicates, are ductile and capable of acquiring strong foliations. With increasing grade, the sheet silicates react to form epidote, amphiboles and plagioclase, and then pyroxenes and possibly garnet. Because of this mineralogical progression, basaltic rocks tend to become less foliated and less ductile at higher grades, as also do pelitic rocks. While perhaps not so colorful as their pelitic associates, mafic rocks nonetheless have mineral assemblages that change distinctively with metamorphic grade and so are valuable grade indicators in the field (Fig. 1.1).

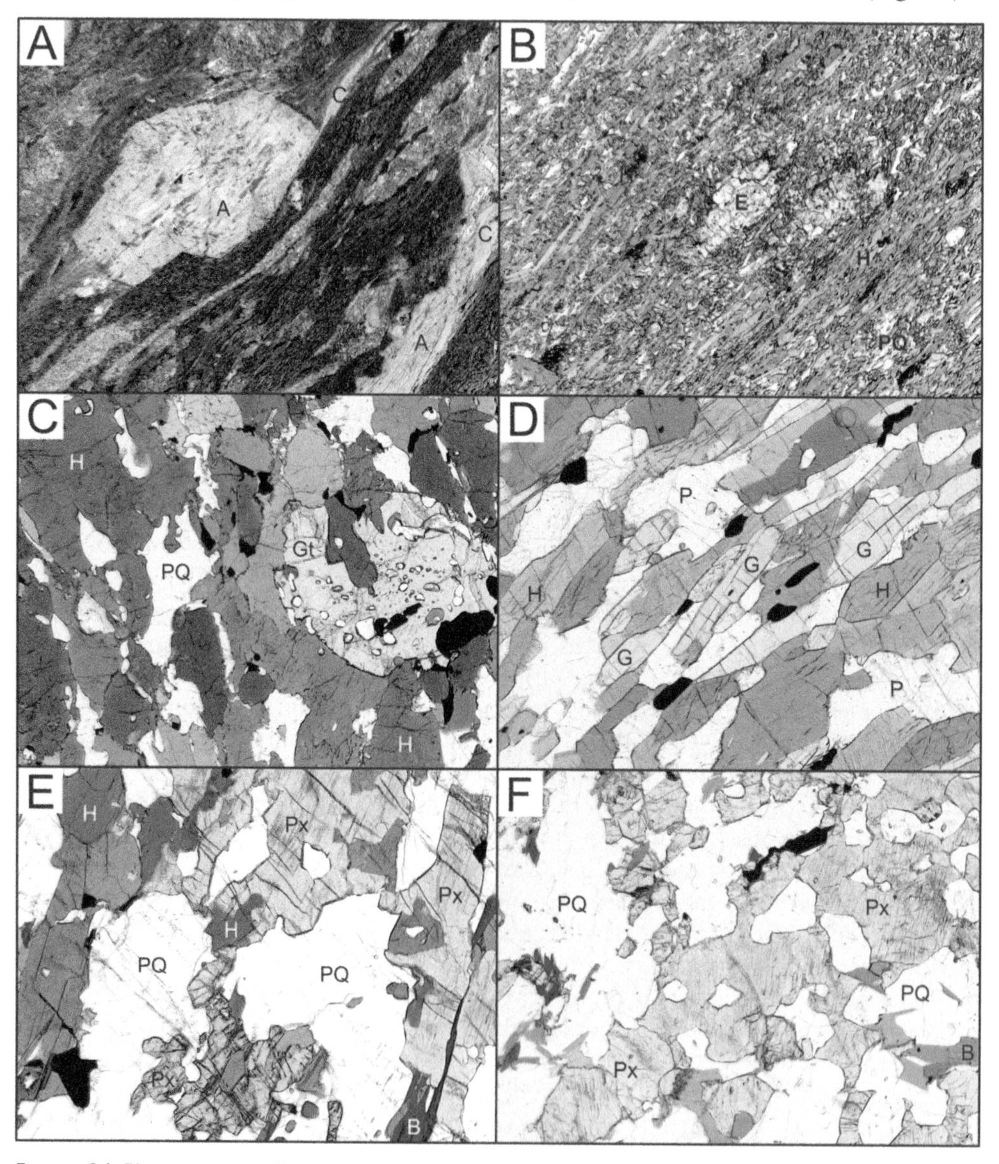

Figure 9.1 Photomicrographs of mafic rocks in thin section, showing mineralogy and textures with increasing metamorphic grade. All are in plane polarized light and have field widths of 4 mm. A) Greenschist facies, with the assemblage chlorite – calcite – white mica – actinolite. The large, apparently euhedral actinolite crystal is a pseudomorph, having replaced an igneous augite phenocryst. Littleton, New Hampshire, USA. B) Epidote amphibolite facies, with the assemblage hornblende – epidote – chlorite – plagioclase – calcite – quartz. Kvithylla, outer Trondheimsfjord, Sør Trøndelag, Norway. C) Amphibolite facies. This relatively iron-rich rock has the assemblage hornblende – garnet – plagioclase – quartz. Quabbin Reservoir, Massachusetts, USA. D) Amphibolite

Figure 9.1 (Continued) facies, to contrast with C. This relatively magnesium-rich rock has the somewhat different assemblage hornblende – gedrite – plagioclase (no quartz). Mafic rocks can have a relatively broad range of Mg/Fe ratios, so it is important to remember that composition, in addition to the metamorphic conditions, can affect mineralogy. North Orange, Massachusetts, USA. E) Transition between the amphibolite and granulite facies, with the assemblage hornblende – augite – enstatite – plagioclase – quartz – biotite. Warren, Massachusetts, USA. F) Granulite facies, with the assemblage augite – enstatite – plagioclase – quartz – biotite. Union, Connecticut, USA. There is a general progression of increasing grain size, and less prominent foliation, with increasing grade. Larger grain size is promoted by higher temperatures, which hastens recrystallization and grain growth. Foliation becomes less pronounced because of the loss of sheet silicates such as chlorite and white micas, and eventually hornblende and some biotite. Abbreviations: A, actinolite; B, biotite; C, chlorite; E, epidote; G, gedrite; Gt, garnet; H, hornblende; P, plagioclase; PQ, plagioclase and quartz; Px, pyroxene.

Mafic rock units such as metamorphosed lavas can be important marker horizons for outlining geologic structures. Their geochemistry can be useful for constraining the paleotectonic environments in which the rocks were originally emplaced. Pillow lavas unmistakably indicate an underwater extrusive setting (if they can be unmistakably identified), and extensive dike sets indicate an extensional setting. Metamorphosed basaltic rocks can therefore be a valuable focus of multi-faceted field and laboratory investigations, despite the discomfort they may have caused earlier generations of geologists (i.e., the quote above).

The thin section images of metamorphosed mafic rocks (Fig. 9.1) are arranged in order of increasing metamorphic grade, from greenschist to pyroxene granulite facies. The field photos are also arranged in order of increasing metamorphic grade: greenschist facies (Figs. 9.2–9.4), epidote amphibolite facies (Figs. 9.5–9.9), amphibolite facies

Figure 9.2 Basaltic pillow lava metamorphosed to greenschist facies. The pillows have been deformed very little, so their shapes are relatively intact. Surface staining causes the apparent rock color to vary over the outcrop surface (where unstained it is gray-green), but the pillow shapes are distinct (one outlined in red). The yellow arrow points to a small breccia deposit, probably made of broken material that slid off the lava flow front, covered soon after by more pillows. The pillows have bulging tops and cuspate bottoms, where the newly-forming, plastic pillows flowed into spaces between underlying pillows. These features indicate that the stratigraphic top direction is to the upper right. Sonoma County coast, California, USA.

(9.10–9.14), and pyroxene granulite facies (9.15–9.17). The field photos include examples of both mafic rocks that have experienced severe deformation, and those that have not for reasons of anhydrous mineralogy or tectonic happenstance.

Figure 9.3 Vertical dikes, little deformed, metamorphosed to greenschist facies. Yellow arrows indicate dike widths. The V-shaped dike fragment, to the right, was apparently a vertical dike that was truncated by the left-dipping dike that is partly behind the person. In this ophiolite complex, the metamorphic grade varies with depth from amphibolite facies in the deeper gabbros to prehnite-pumpellyite facies in the shallower pillow lavas. Metamorphism was associated with ocean crust (probably forearc, Bédard, 1999) hydrothermal activity at the time of crust formation, rather than by contact or regional metamorphism. Betts Cove area, Newfoundland, Canada.

Figure 9.4 Folded basaltic rocks metamorphosed to upper greenschist facies conditions. Notice that the rocks in the distance, even those at the distant shoreline (red arrow), weather brownish-gray instead of grayish-green, indicating a different rock type. Contacts can be mapped by such intermediate-range, eyeball-type remote sensing, without having to walk every bit of the contact. Baie Verte, Newfoundland, Canada.

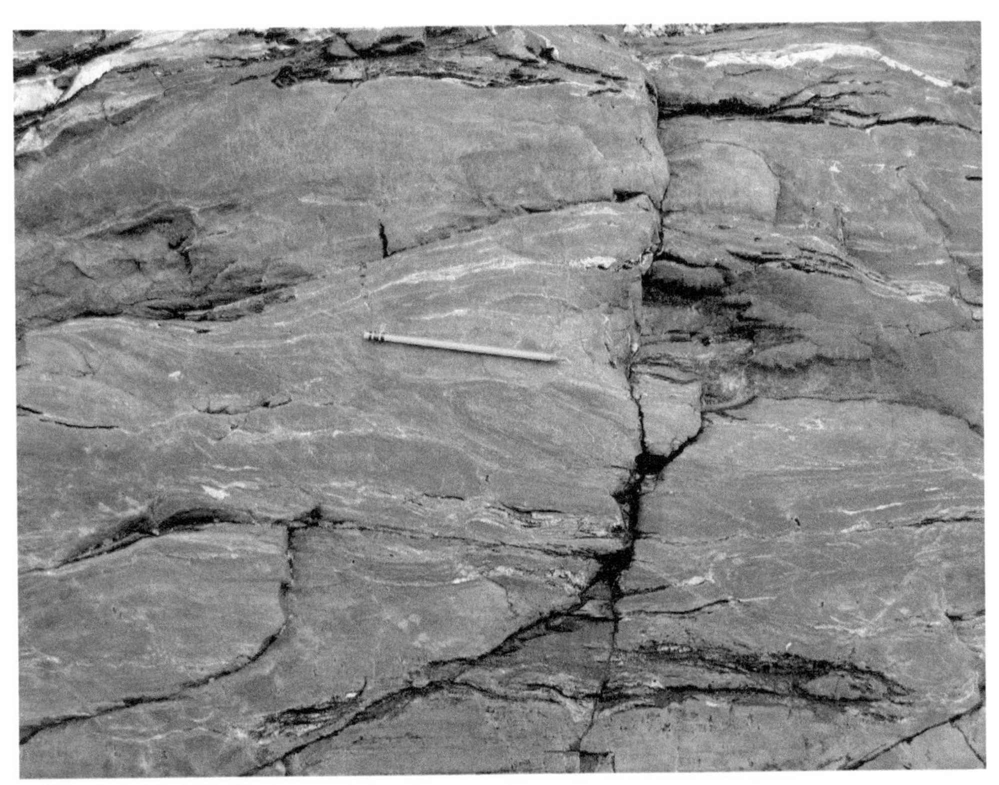

Figure 9.5 Boudinaged basaltic layers metamorphosed to epidote amphibolite facies, on a recently exposed, glacially polished surface. The greenish color of the rocks is imparted by epidote, relatively light-green hornblende, and chlorite. The layering is approximately parallel to the pencil, and one of many small boudins can be seen just below and to the right of the pencil. While not diagnostic, small-scale inhomogeneities, irregular, thin quartz veins, and biotite-rich selvages suggest that this was originally a series of lava flows or dikes. Mafic rocks having gabbroic protoliths, in this area, tend to be coarser-grained and more homogeneous. Rekdal, outer Moldefjord, Møre og Romsdal, Norway.

Figure 9.6 Deformed, hydrothermally altered mafic rocks at epidote-amphibolite facies. This outcrop is composed of epidote-bearing vein quartz boudins in a matrix of fine-grained, dark, chlorite – hornblende phyllite. The white to light-green epidote – quartz boudins are interpreted to have been hydrothermal vein deposits in fractures in the host rock. This is probably part of or similar to the hydrothermal system that produced the nearby magnetite-sulfide vein shown Figure 19.17. Esvikneset, outer Trondheimsfjord, Sør Trøndelag, Norway.

Figure 9.7 Gabbro, metamorphosed at epidote amphibolite facies, that is almost undeformed, shown wet and dry on a freshly broken surface. It contains the mineral assemblage chlorite – hornblende – epidote – calcite – albite – quartz, and is much coarser than nearby metamorphosed basalts. In thin section, relict augite can be found in the cores of some hornblende pseudomorphs after augite. Støräs, Sør Trøndelag, Norway.

Figure 9.8 Metamorphosed and deformed gabbro at epidote amphibolite facies. This rock is much coarser-grained than other nearby basaltic rocks, that generally look like that in Figure 9.5. The hornblende porphyroclasts may be pseudomorphs after augite, as indicated by augite cores in some large hornblende crystals in metamorphosed gabbro at somewhat lower grade (Fig. 9.7). The protolith for this rock was probably a cumulate gabbro, based on high Ca, Al, and Mg concentrations, and very low incompatible element concentrations (e.g., Th, Zr, La) compared to some other coarse-grained, gabbroic-looking rocks, and all of the local finer-grained amphibolites. Bolsøy, Moldefjord, Møre og Romsdal, Norway.

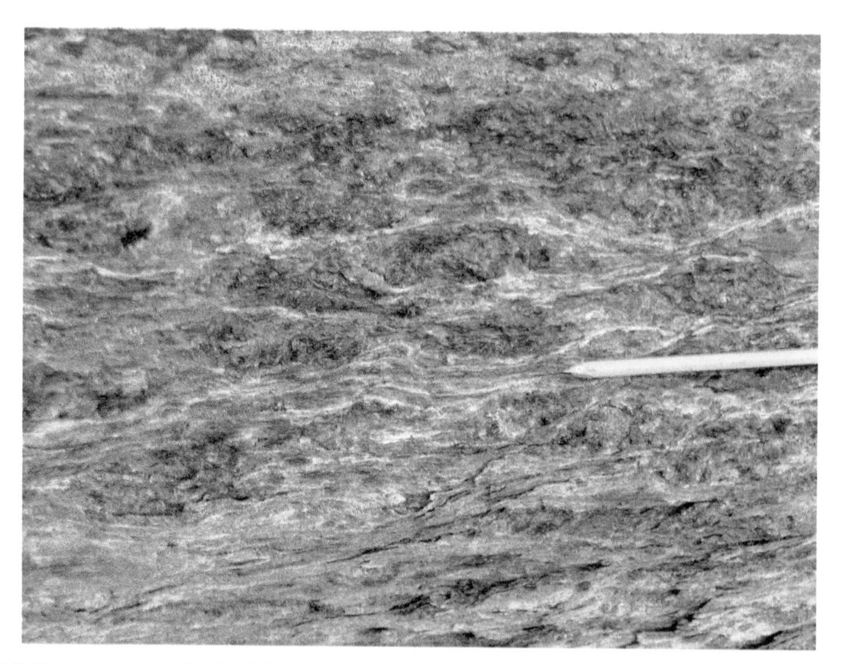

Figure 9.9 Very coarse-grained gabbro metamorphosed to epidote amphibolite facies conditions. This rock has hornblende porphyroclasts up to 8 cm long in a matrix of hornblende, plagioclase, epidote, and minor biotite. The hornblende contains abundant plagioclase inclusions, suggesting that the hornblende may originally have been augite oikocrysts. Litleneset, outer Trondheimsfjord, Sør Trøndelag, Norway.

Figure 9.10 Garnet amphibolite metamorphosed to amphibolite facies conditions. This rock has a weak hornblende lineation that is approximately parallel to the red line. The pink garnet porphyroblasts seem to be mostly on the left side of the image, indicating that the two sides have slightly different chemical compositions. It is unclear if the protolith for this rock was gabbroic or basaltic. Bolsøy, Moldefjord, Norway.

Figure 9.11 Deformed dark-gray basaltic dikes that intruded lighter-gray granitic gneiss, then all metamorphosed to amphibolite facies conditions. These once-continuous dikes intruded during an episode of continental rifting. They were later folded, boudinaged, and cut by small white quartz veins during metamorphism. While too highly deformed here to easily see that the amphibolites were dikes, elsewhere it is clear that the amphibolites cut gneissic layering (Fig. 16.11). Lepsøy, Nordøyane, Møre og Romsdal, Norway.

Figure 9.12 Fine-grained amphibolite metamorphosed to amphibolite facies conditions, with elongate, white, monomineralic spots of plagioclase. These are interpreted to be recrystallized plagioclase phenocrysts, based on their pure plagioclase mineralogy, and the nearby presence of apparently relict igneous plagioclase cores in some particularly large plagioclase porphyroclasts (Fig. 25.1C, D). Quabbin Reservoir, Massachusetts, USA.

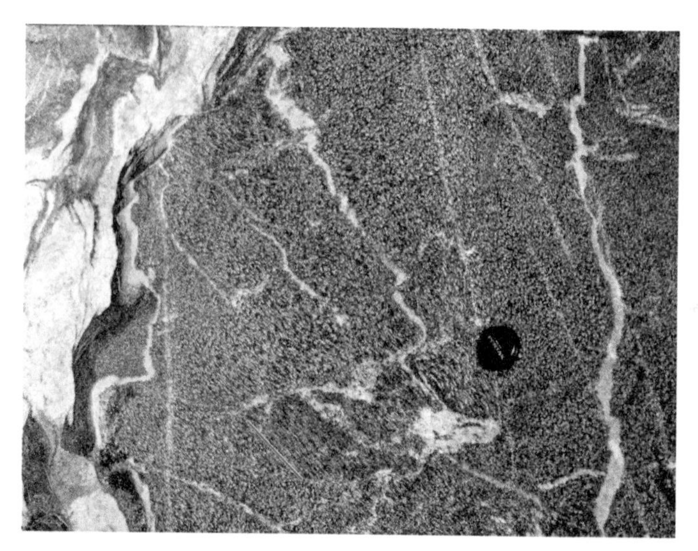

Figure 9.13 This is a boudin of coarse-grained, dark-gray basaltic rock, possibly originally a gabbro, hosted in white and gray tonalitic gneiss (left side only). These were metamorphosed at amphibolite facies conditions. The tentative gabbro interpretation is based on the fact that this amphibolite is part of a group of large boudins that are much coarser than most other amphibolites in the area (e.g., Fig. 9.12). Most of the amphibolite surface has spots of white plagioclase and black hornblende. In between the red line and the lens cap is a strong mineral lineation, parallel to the red line. The spots and lines are probably the same lineation, folded so it can be seen in both end-section and length-section, respectively. The amphibolite and surrounding gray felsic gneiss have been cut by white, relatively fine-grained tonalitic dikes that formed during partial melting of the host gneiss. Quabbin Reservoir, Massachusetts, USA.

Figure 9.14 Pyroxene-bearing amphibolite that has been metamorphosed to conditions transitional between amphibolite and granulite facies. The dark-gray, angular blocks have the assemblage hornblende – plagioclase – enstatite – augite. The lighter, brownish rock has almost the same assemblage as the blocks, but it is quartz-rich and has more plagioclase and less pyroxene and hornblende, making the light-brown rocks tonalitic. The light brown rocks are interpreted to be melt segregations from the dark-gray rock. This inference was made because they contain the same assemblage, except for quartz, as though the melting reaction proceeded to the exhaustion

Figure 9.14 (Continued) of quartz. The grain boundary liquid then separated into fractures, where it crystallized, leaving behind a solid, quartz-free residue. The liquid probably moved no more than several centimeters or meters. The brown spots (red arrows) are particularly large, weathered orthopyroxene crystals, and the brown lines (yellow arrows) are late dolomite-filled fractures. Fort Ann, Adirondacks, New York, USA.

Figure 9.15 Coarse basaltic dike (bottom half) cutting anorthosite (top half), metamorphosed at granulite facies conditions. The dike assemblage is plagioclase – augite – enstatite – magnetite – garnet. Most of the red garnet formed at pyroxene – plagioclase and magnetite – plagioclase contacts, where solid state diffusion permitted garnet to grow from the chemical components available in adjacent minerals. Amazingly, some of the plagioclase is still gray, containing micron-scale Fe-Ti oxides that exsolved from the original igneous plagioclase solid solution. The width of the field of view is about 20 cm, so the coarser crystals are more than a centimeter across. Jay, Adirondacks, New York, USA.

Figure 9.16 Ferrobasaltic dike cutting anorthosite (not shown), metamorphosed at granulite facies conditions. This dike contains a very large proportion of metamorphic garnet and pyroxene, about

Figure 9.16 (Continued) 30% each. The lath-like shapes of the original plagioclase phenocrysts are still visible, though most of the plagioclase has recrystallized. The large, gray plagioclase crystal to the lower right is probably a xenocryst from the anorthosite. It has not recrystallized and still contains abundant, exsolved Fe-Ti oxides that cause the gray color. The rock is little-deformed, probably because it was emplaced into strong, dry anorthosite which itself resisted deformation. Elizabethtown, Adirondack Mountains, New York, USA.

Figure 9.17 Though perhaps not quite a mafic rock, anorthosite is generally related to mafic magmas deep in the crust, probably forming from them as plagioclase-rich cumulates. This is an example of coarse anorthosite that has been metamorphosed at granulite facies conditions. The gray porphyroclasts are relict igneous plagioclase crystals, gray because of micron-scale exsolved Fe-Ti oxides. The oxides precipitated from the original Fe- and Ti-bearing igneous plagioclase solid solution. The white matrix is recrystallized plagioclase, having grain sizes generally 0.1-1 mm, in which the micron-scale oxide inclusions have recrystallized into coarser grains that are less able to color the rock. Speculator, Adirondacks, New York, USA.

Chapter 10

Blueschists

Blueschist facies rocks form at low temperature and relatively high pressure conditions (Fig. 1.1) that are only found in subduction zones. Blueschists in orogenic belts are evidence of former subduction zones, as are their higher temperature cousins, orogenic eclogites. Basaltic and other feldspathic rocks metamorphosed in the blueschist facies fields undergo reactions that form the blue amphibole glaucophane. Because low metamorphic temperatures tend to limit grain growth rates and hence grain size, glaucophane grain sizes are typically small. Small grain size, combined with the elongate and slightly flattened crystal shape typical of amphiboles, plus maybe a little white mica or chlorite, allows the rocks to be somewhat schistose with notable foliations and mineral lineations.

The type mineral of blueschists, glaucophane, forms at high pressure from the albite component of plagioclase, or the equivalent in paragonite in some cases. At blueschist facies glaucophane is the principal host of sodium, as omphacite is at eclogite facies. Example reactions for making glaucophane include:

From a dry basaltic rock during infiltration of aqueous fluid:

$$\underset{\text{albite}}{2NaAlSi_3O_8} + \underset{\text{enstatite}}{3MgSiO_3} + \underset{\text{fluid}}{H_2O} = \underset{\text{glaucophane}}{2Na_2Mg_3Al_2Si_8O_{22}(OH)_2} + \underset{\text{quartz.}}{SiO_2} \quad (10.1)$$

$$\underset{\text{albite}}{4NaAlSi_3O_8} + \underset{\text{olivine}}{3Mg_2SiO_4} + \underset{\text{silica*}}{SiO_2} + \underset{\text{fluid}}{2H_2O} = \underset{\text{glaucophane.}}{2Na_2Mg_3Al_2Si_8O_{22}(OH)_2} \quad (10.2)$$

*From reaction 10.1, for example, because these and other reactions can take place concurrently. Here are some example reactions for making glaucophane from a hydrated, lower grade rock:

$$\underset{\text{albite}}{4NaAlSi_3O_8} + \underset{\text{chlorite}}{3Mg_5Al_2Si_3O_{10}(OH)_8} + \underset{\text{quartz}}{SiO_2} =$$
$$\underset{\text{glaucophane}}{2Na_2Mg_3Al_2Si_8O_{22}(OH)_2} + \underset{\text{garnet}}{3Mg_3Al_2Si_3O_{12}} + \underset{\text{fluid.}}{10H_2O} \quad (10.3)$$

$$\underset{\text{albite}}{10NaAlSi_3O_8} + \underset{\text{actinolite}}{3Ca_2Mg_5Si_8O_{22}(OH)_2} + \underset{\text{lawsonite}}{3CaAl_2Si_2O_7(OH)_2 \bullet H_2O} =$$
$$\underset{\text{glaucophane}}{5Na_2Mg_3Al_2Si_8O_{22}(OH)_2} + \underset{\text{garnet}}{3Ca_3Al_2Si_3O_{12}} + \underset{\text{quartz}}{11SiO_2} + \underset{\text{fluid.}}{4H_2O} \quad (10.4)$$

In these reactions, Mg-olivine, albite, and the other end members are not actually pure minerals in the real rocks, but instead are solid solutions. For example, Mg in the reactions refers to both Mg and Fe^{2+} in solid solutions. Also, albite in the real rock is part of the plagioclase Na-Ca solid solution series, but it is the Na (albite) part that is involved in the four reactions above. The Ca component of plagioclase, anorthite, is involved in other reactions, such as forming the grossular component of garnet. The grossular component (Ca-garnet) combines with the Fe-Mg garnet components to make Ca-Fe-Mg garnet solid solutions. The reactions here use pure end members because they are easier to follow, an approximation that gets the point across.

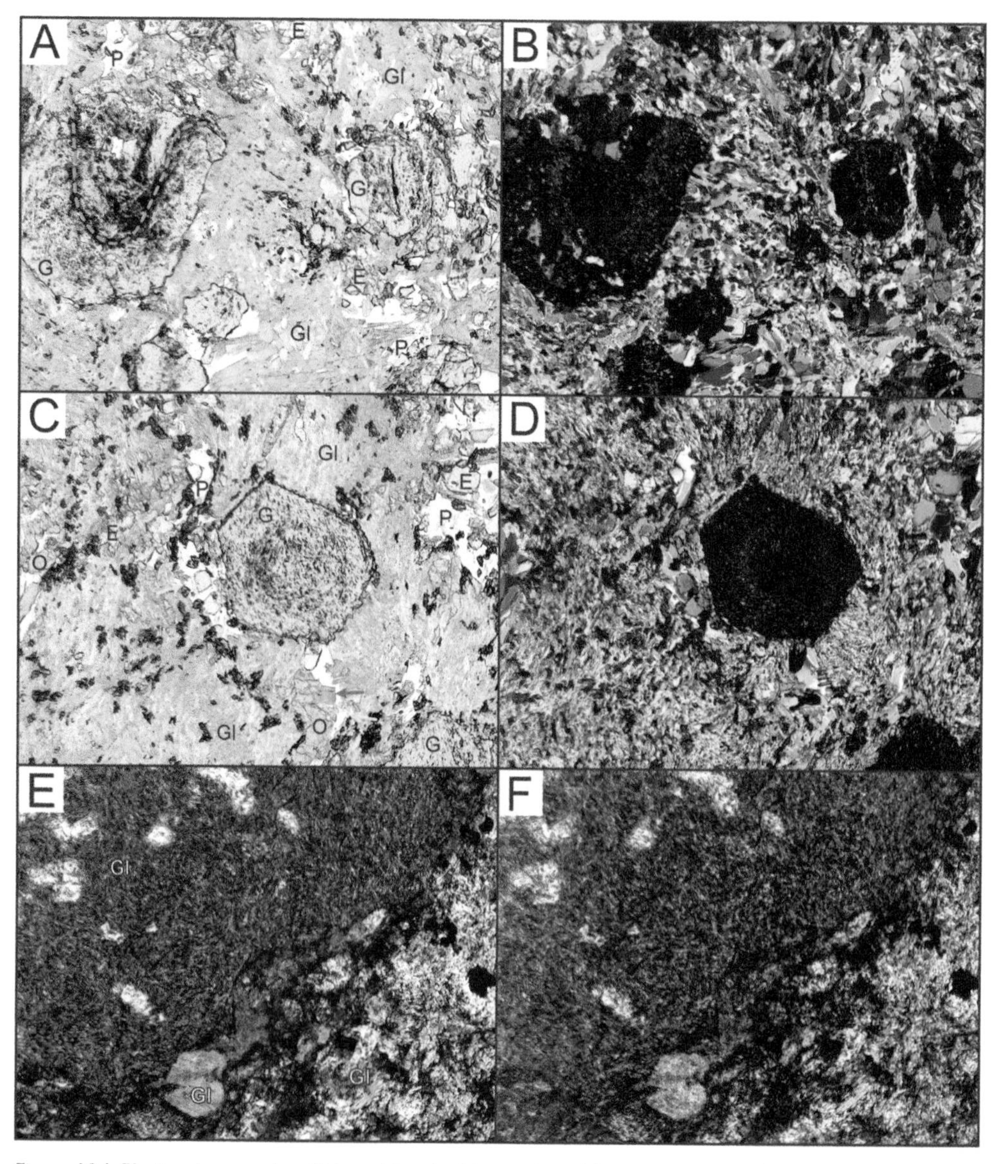

Figure 10.1 Photomicrographs of blueschists in thin section. All images on the left are in plane polarized light, and those on the right are the same fields in cross-polarized light. All field widths are 4 mm. A, B) Blueschist with the assemblage glaucophane – garnet – epidote – phengite. The dark, cloudy material

Figure 10.1 (Continued) is fine-grained titanite. Notice how the garnet has overgrown an earlier folded foliation, indicated by the blue dashed line in A. Though not visible here, some garnets contain green hornblende in their cores, recording an earlier, lower pressure part of the metamorphic history. C, D) Blueschist that is transitional to eclogite facies, having the assemblage glaucophane – garnet – omphacite – epidote – phengite. The P-T conditions for the transition from blueschist to eclogite depends on both and fluid and rock compositions. Extending from the left side of the center garnet, toward the upper left and lower right, is a vein (yellow arrows) containing phengite, epidote, minor quartz, and glaucophane that has concentric color zoning (red arrow). These veins were deposited by flowing fluid, that appears as a network of irregular, gray to white lines on the outcrop surface. E, F) This rock is a volcanic breccia containing basaltic clasts (upper left) in a lighter-colored fragmental matrix (lower right). While at first glance images E and F (and the rock itself) look like a fine-grained mess, look carefully and you can see the texture of randomly-oriented lath-shaped crystals in the basalt fragment. In basalt the laths were plagioclase, but now they have been replaced by blue glaucophane. Somewhat larger glaucophane crystals are visible in the matrix, along with epidote and phengite. In the field, the minor amount of glaucophane gave the whole rock a blue-gray appearance, subtly contrasting it with other glaucophane-free rocks. Abbreviations: E, epidote; G, garnet; Gl, glaucophane; O, omphacite; P, phengite.

Extensive blueschist terranes are found in orogenic belts, but blueschists and related rocks can also be restricted to isolated blocks in rocks of generally lower metamorphic grade (a mélange). Such disruption and mixing of different rock types may occur during deformational reworking in the subduction zone accretionary wedge, or collapse of tectonically exposed rocks from the margin of the accretionary wedge into deep water of the trench. Isolated blocks of blueschist facies rocks in lower grade materials tend to weather out in raised relief, allowing them to stand out on landscape surfaces.

Figure 10.2 Typical blueschist on a wave-polished surface. The rock consists of red garnet porphyroblasts in a matrix of fine-grained glaucophane. With a hand lens, lighter-colored grains of epidote and phengite can be seen. If the sun is shining, the glinting, highly reflective phengite cleavage surfaces are brilliantly visible. Sonoma County coast, California, USA.

The thin section photos (Fig. 10.1) show one blueschist sample (A, B), and another that is transitional from blueschist to eclogite facies because of the presence of omphacite, a pyroxene characteristic of eclogites (C, D). The last sample is a fine-grained, essentially undeformed rock that has minor glaucophane with a variety of other minerals (E, F). This rock is special first because it has the blueschist mineral assemblage despite being almost undeformed. Second, it was picked out of a river gravel bar because it appeared bluish-gray, contrasting with the more common ordinary gray rocks. The small amount of glaucophane made the rock distinctive. Moral: train your eyes to see things that are different. The field photos are arranged to show glaucophane and glaucophane – phengite schists (Figs. 10.2, 10.3), glaucophane schist that is transitional to eclogite (Figs. 10.4, 10.5), and glaucophane schist interlayered with eclogite (Figs. 10.6, 10.7). Interlayering of eclogite and blueschist requires that the two rocks have different chemical compositions, with the eclogite layers possibly being more Mn- or Fe-rich (Oh et al., 1991).

Figure 10.3 Glaucophane-bearing rock (right and upper left) gradational into rusty-weathering quartz – phengite rock (left). The rusty-weathering rock contains small amounts of sulfide, which weathers to produce the brownish limonite stain. Close examination of the rusty rock with a hand lens shows that it contains small amounts of glaucophane in addition to quartz and mica, and so records the same blueschist facies metamorphic grade as the surrounding rock. The protolith for this outcrop may have been an interlayered, variable mixture between silty or shaley sediment and basaltic volcaniclastics. Notice the tight fold hinges to the right of the hammer pick. Sonoma County coast, California, USA.

Figure 10.4 This omphacite-bearing glaucophane schist is in sharp contact with a layer of mica schist (upper right). The schist is composed mostly of phengite, garnet, quartz, and epidote, with traces of glaucophane. The blue glaucophane stands out beautifully against the white mica, as seen with a hand lens. The phengite schist is probably a metamorphosed shale that was deposited in contact with the original basalt. Sonoma County coast, California, USA.

Figure 10.5 Blueschist broken along the dominant cleavage surface, showing small, sub-parallel folds on the surface that have axes approximately parallel to the red line. These are a weak crenulation cleavage that cuts the earlier cleavage, forming a cleavage-cleavage intersection lineation (see Chapter 15). This rock contains green omphacite in addition to garnet, so it is transitional to eclogite facies. The field width is about 7 cm. Sonoma County coast, California, USA.

Figure 10.6 Interlayered glaucophane schist and eclogite, showing how compositional differences can affect mineral proportions and grain sizes. Below and to the right of the knife is a layer of glaucophane-rich rock that has large garnets that make up about 10% of the layer (red arrow). A similar layer is also visible at the bottom of the image (dashed red arrow). The layers indicated by yellow arrows are glaucophane schist containing much smaller garnets that make up about 5% of each layer. These are sandwiched between layers and irregular masses of green eclogite and layers intermediate between blueschist and eclogite. Outcrops like this are good places to collect samples for phase relation studies, because a lot of compositional variation is present in a small area at one metamorphic grade. Sonoma County coast, California, USA.

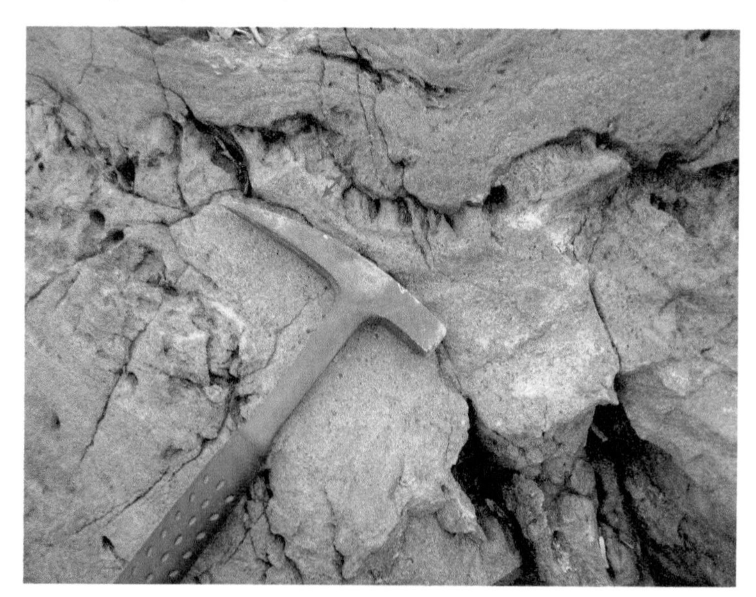

Figure 10.7 Green eclogite layer (red arrow) in sharp contact with blueschist (above) and light-green to light-blue phengite-rich rock (center). The glaucophane-bearing rock to the lower left and lower right seem to be gradational mixtures between basaltic and the phengite-rich rock compositions. The relatively homogeneous blueschist and eclogite layers may represent lava flows or a dikes, but the rest is probably a range of volcaniclastic-shale mixtures. They all clearly have different compositions. Sonoma County coast, California, USA.

Chapter 11

Eclogites

Eclogites are high-pressure metamorphosed basalts, characteristically composed of about half red Ca-rich garnet and half green omphacite, an Na-Ca clinopyroxene intermediate between augite and jadeite. They form at temperatures and usually pressures higher than blueschists, but typically lower than those expected along a typical continental geotherm. Because of the unusual combination of relatively low temperature and high pressure, eclogites are typically found in fossil subduction zones where cold basaltic rocks were brought to great depths, then returned to the surface fast enough for the eclogite mineralogy to have been preserved. Eclogites can also be brought to the surface from the deep continental lithosphere as xenoliths in some alkali basalts and kimberlites. Typically, orogenic eclogites are found as boudins ranging from fist-size to hundreds of meters across. Eclogites can also occur as enormous masses or tectonic slabs that are kilometers across.

The unique garnet – omphacite mineralogy of fresh eclogite forms as a result of plagioclase reacting at high pressure to form denser assemblages:

$$\underset{\text{albite}}{NaAlSi_3O_8} = \underset{\text{jadeite}}{NaAlSi_2O_6} + \underset{\text{silica.}}{SiO_2} \tag{11.1}$$

$$\underset{\text{anorthite}}{3CaAl_2Si_2O_8} = \underset{\text{grossular}}{Ca_3Al_2Si_3O_{12}} + \underset{\text{kyanite}}{2Al_2SiO_5} + \underset{\text{silica.}}{SiO_2} \tag{11.2}$$

For ease of reading the reactions, the albite and anorthite components of plagioclase solid solutions have been separated. The jadeite component forms a solid solution with augite (from augite in the protolith or from the same chemical components in amphiboles or other minerals) to form omphacite:

$$\underset{\text{jadeite}}{xNaAlSi_2O_6} + \underset{\text{augite}}{yCaMgSi_2O_6} = \underset{\text{omphacite.}}{x{+}y(Na,Ca)(Al,Mg)Si_2O_6} \tag{11.3}$$

As another simplification, Mg in these reactions refers to Mg and Fe^{2+} in the actual rocks. Kyanite from reaction 11.2 does not actually occur in most eclogites, because the excess Al is usually consumed by garnet-forming reactions. Next, reaction 11.2 has been re-written to include an igneous Fe-Mg mineral, orthopyroxene or olivine, that eliminates kyanite from the reaction products:

$$\underset{\text{anorthite}}{3CaAl_2Si_2O_8} + \underset{\text{orthopyroxene}}{2MgSiO_3} = \underset{\text{Ca garnet}}{Ca_3Al_2Si_3O_{12}} + \underset{\text{Mg-Fe garnet}}{2Mg_3Al_2Si_3O_{12}} + \underset{\text{quartz.}}{SiO_2} \tag{11.4}$$

$$3CaAl_2Si_2O_8 + MgSiO_3 = Ca_3Al_2Si_3O_{12} + 2Mg_3Al_2Si_3O_{12} \quad (11.5)$$

anorthite olivine Ca garnet Mg-Fe garnet.

In real rocks these reactions might take place simultaneously. As with plagioclase and omphacite, the Ca and Mg-Fe garnet components in the simplified reactions above form their own one-garnet solid solution series:

$$xCa_3Al2Si_3O_{12} + y(Mg,Fe)_3Al_2Si_3O_{12} = x+y(Ca,Mg,Fe)_3Al_2Si_3O_{12} \quad (11.6)$$

grossular Mg-Fe garnet Ca-Mg-Fe garnet.

The result, in a fresh eclogite, is a spectacular rock containing green pyroxene and red garnet. Because the eclogite mineral assemblage is not stable at low pressures, and because they form at higher temperatures than blueschists, eclogites are commonly retrograded en route to the surface. They usually retrograde toward medium-pressure assemblages that can range from greenschist to pyroxene granulite facies. Retrograde metamorphism can take place at different scales, from thin grain contact regions to entire large eclogite masses. Figure 11.1 shows a typical example of an eclogite boudin that preserves the eclogite mineral assemblage only in the boudin core. The eclogite rim has been retrograded to amphibolite, having different mineralogy but almost the same chemical composition as the eclogite core. The boudin is characteristically surrounded by seemingly unremarkable, medium-pressure metamorphic rock (e.g., muscovite schist, granitic gneiss, quartzite). Clearly the eclogite core, rim, and at least the immediately adjacent host rock must have experienced the same eclogite facies metamorphic conditions, but the latter two have reequilibrated during exhumation.

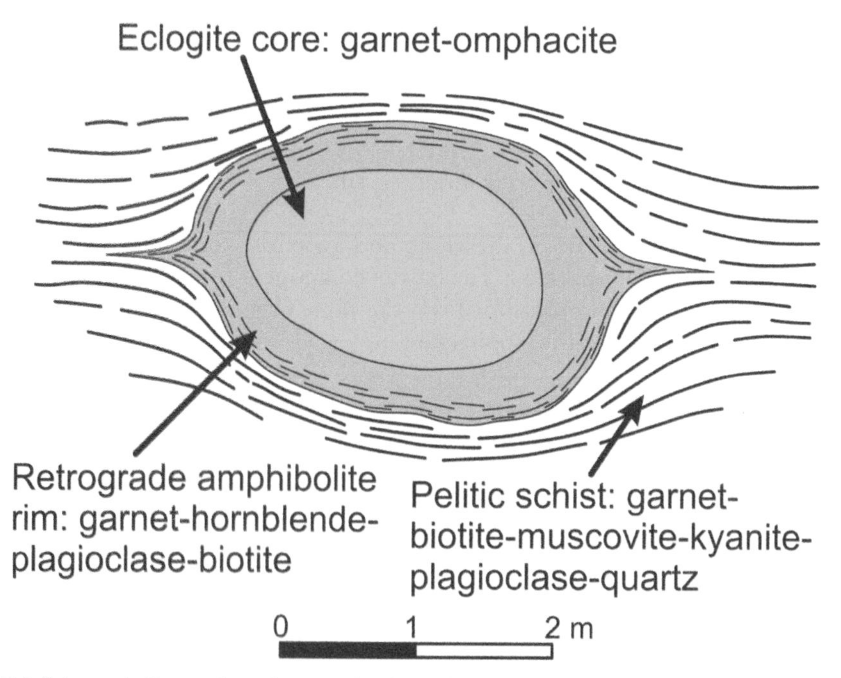

Figure 11.1 Schematic illustration of a typical eclogite boudin, containing eclogite in the boudin core, with a rim of eclogite that has been retrograded to lower pressure conditions (here to amphibolite facies). The boudin is surrounded by host rock, also at amphibolite facies (modified after Peacock and Goodge, 1995).

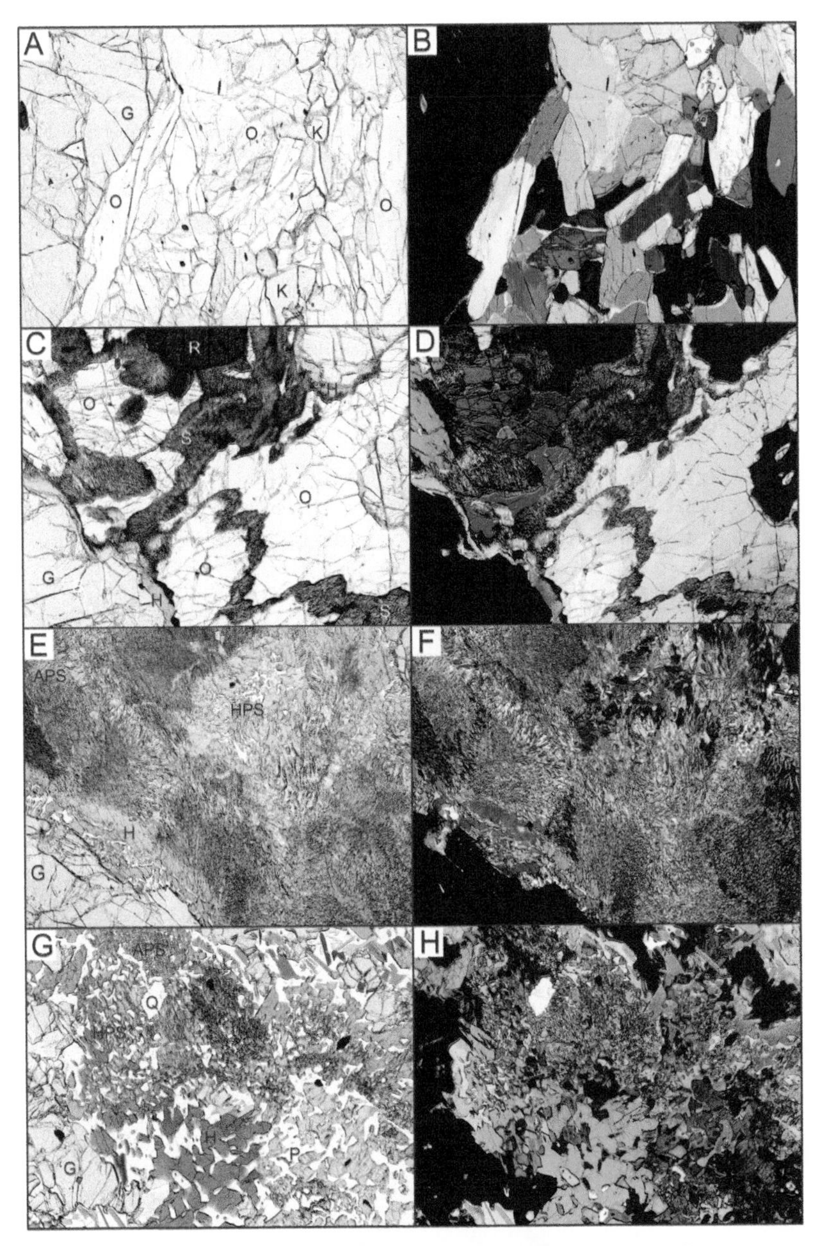

Figure 11.2 Photomicrographs of fresh eclogite, and eclogite that has been retrograded to varying degrees to amphibolite facies mineral assemblages. Left images are in plane-polarized light, and right images are the same fields in cross-polarized light. All field widths are 4 mm. A, B) Fresh kyanite eclogite, with the assemblage omphacite – garnet – kyanite. C, D) Partially retrograded eclogite with about 50% of the omphacite and 95% of the garnet remaining. Garnet is rimmed by green hornblende, and the fine-grained augite – plagioclase symplectite after omphacite appears dark-gray. E, F) Retrograded eclogite with no omphacite and about 80% of the garnet remaining. Garnet is surrounded by a plagioclase – hornblende symplectite, and the omphacite has been replaced by hornblende – plagioclase – augite symplectite. G, H) Like E and F, but after recrystallization and coarsening of the symplectites. Recrystallization may progress to the point where there is little or no evidence that the rock was ever an eclogite (compare to Fig. 9.1C). Abbreviations: G, garnet; H, hornblende; HPS, hornblende – plagioclase symplectite; K, kyanite; O, omphacite; P, plagioclase; Q, quartz; R, rutile; S, pyroxene – plagioclase symplectite.

On a smaller scale, retrograde metamorphism of the eclogite mineral assemblage commonly begins with decomposition of omphacite into a symplectite (wormy intergrowth) of augite and plagioclase, and in some cases by decomposition of Ca-rich garnet into a symplectite of plagioclase and pyroxene (Fig. 11.2). If aqueous fluid is available, hornblende or other hydrous minerals can replace the pyroxenes and some plagioclase. In the field, omphacite in fresh eclogite is usually relatively transparent and pale- to medium-green. As the eclogite becomes retrograded, the original omphacite turns into dull, gray-green symplectite, and even black if hornblende becomes abundant.

The thin section photos show progressive retrograding of fresh eclogite to an amphibolite facies assemblage that retains the symplectite texture, derived from decomposition of high-pressure omphacite and garnet. The recognition of fresh vs. retrograded eclogite can be important if specimen collecting is for mineral assemblages rather than, say, whole rock chemical analyses. The field photos first show large blocks of eclogite in host rock (Figs. 11.3, 11.4), then meter-scale views of fresh and retrograded eclogite (Figs. 11.5–11.8). Following those are close-up views of fresh (Figs. 11.9–11.11), somewhat retrograded (Fig. 11.12), and highly retrograded (Fig. 11.13) eclogite. Figure 11.14 is a granitic gneiss that was metamorphosed at eclogite facies conditions, and survived without significant retrograding to reach the surface.

Figure 11.3 This is typical of many eclogite occurrences, where eclogite boudins (blue arrows) seemingly swim in an endless sea of gneiss. Here, the boudin margins are retrograded, transformed to varying degrees to amphibolite, representing the same metamorphic facies as the adjacent granitic gneiss. Only the boudin cores (red arrows) retain actual eclogite. Assuming boudins generally don't escape from great depths like slippery melon seeds from between squeezed fingers, the eclogites and host gneisses must have experienced similar metamorphic conditions. Ullaholmen, Nordøyane, Møre og Romsdal, Norway.

Figure 11.4 Three large, angular boudins of eclogite (dark-brown, middle ground, red arrows), surrounded by granitic gneiss (light-gray) and pegmatite (white). Flemsøy, Nordøyane, Møre og Romsdal, Norway.

Figure 11.5 Fresh eclogite, with red garnet, transparent green omphacite, white zoisite, and small amounts of blue kyanite. The protolith of this rock was basaltic, like eclogites in general, but was quite inhomogeneous. Verpeneset, Sogn og Fjordane, Norway.

Figure 11.6 Partially retrograded eclogite having alternating, mostly fresh garnet-rich (reddish) and omphacite-rich (greenish) layers. Severely retrograded black layers (red arrows), rich in hornblende, are where aqueous fluid penetrated during uplift. This is part of a large eclogite body that is kilometers across. Averøy, Møre og Romsdal, Norway.

Figure 11.7 Small, blocky boudin of partially retrograded eclogite surrounded by migmatitic gneiss. This eclogite has a black, hornblende-rich retrograde rim, formed by aqueous fluid infiltrating the relatively anhydrous eclogite. The presence of eclogite boudins such as this one implies a once-continuous layer or dike of mafic rock, now broken into fragments. Though more blocky in shape, the boudin core, boudin rim, and host rock relationships are much like those shown in Figure 11.1. Haramsøy, Nordøyane, Møre og Romsdal, Norway.

Figure 11.8 Strongly retrograded boudin of eclogite hosted by layered, migmatitic gneiss. This former eclogite contains abundant hornblende and an unusual amount of garnet compared to numerous other amphibolites in this area. In thin section, the recrystallized and coarsened symplectite texture, from the breakdown of omphacite and Ca-rich garnet, can be seen (as in Figs. 11.2G, H). That texture makes it clear that this was once an eclogite. This boudin is isolated, with no others like it nearby. Lepsøy, Nordøyane, Møre og Romsdal, Norway.

Figure 11.9 Fine-grained, fresh eclogite that has small variations in grain size and mineral abundance. This is a quartz-bearing eclogite and so probably had very little or no olivine in the protolith (reaction 11.5, quartz is white to gray). Sonoma County coast, California, USA.

Figure 11.10 Fresh kyanite eclogite, with light-green omphacite, red garnet, and light-blue kyanite. Kyanite eclogites are commonly derived from plagioclase-rich mafic cumulates, in which the high-pressure breakdown of the aluminous anorthite component of plagioclase can release aluminum in excess of that required to make garnet (reaction 11.2). The width of the field is about 15 cm. Fjørtoft, Nordøyane, Møre og Romsdal, Norway.

Figure 11.11 Fresh orthopyroxene eclogite having green omphacite, red garnet, and brown orthopyroxene. A horizontal orthopyroxene crystal, 5 x 20 mm in size, can be seen just below and to the left of center, with cleavage surfaces reflecting the sunlight. The protolith for this rock was probably an olivine norite, based on the chemical composition, with too little plagioclase to make garnet from all of the original Fe-Mg silicates. Reaction 11.4, if the anorthite component of plagioclase ran out before orthopyroxene, would produce a rock like this. Gossa Island, Møre og Romsdal, Norway.

Figure 11.12 Partially retrograded eclogite. The garnets are surrounded by black, hornblende – plagioclase symplectite rims. The dull, light-green matrix is made mostly of augite – plagioclase symplectite produced from the breakdown of omphacite, along with some black hornblende. The dull reflectivity and lack of transparency of the green material indicates that symplectite has replaced the usually glassy-clear, fresh omphacite. Fjørtoft, Nordøyane, Møre og Romsdal, Norway.

Figure 11.13 Strongly retrograded eclogite, with small, recrystallized red garnets, black hornblende, and dull green areas made of partially recrystallized plagioclase – augite symplectite. Southern Moldefjord, Møre og Romsdal, Norway.

Figure 11.14 This is an example of a rare granitic gneiss that was metamorphosed to eclogite facies conditions, and still retains its eclogite facies assemblage of K-feldspar – quartz – garnet – kyanite. The photo is from a large block or boudin, tens of meters across, hosted by amphibolite facies gneiss. The kyanite is almost white in most of the rock, but the inset shows a small, enlarged area from the same outcrop where the kyanite is more abundant and blue. Fjørtoft, Nordøyane, Møre og Romsdal, Norway.

Chapter 12

Ultramafic rocks

Ultramafic rocks must have more than 90% of mafic minerals, which are Mg- and Fe-rich minerals that include olivine, Mg-, Fe-, and Ca-rich pyroxene, garnet, biotite, chlorite, talc, serpentine, amphiboles, and others. Eclogites, blueschists, and other rocks that have more than 90% of minerals that are arguably mafic (omphacite and glaucophane, respectively, plus garnet) have their own special designations, and are not considered ultramafic. The largest potential source for ultramafic rocks globally is Earth's mantle. Parts of the mantle, typically that associated with oceanic lithosphere, have been brought to observable levels along thrust faults during volcanic arc-continent collisions. Ultramafic rocks of the oceanic lithosphere typically occur as fresh or altered rock in the lower parts of ophiolites. Such rocks can also be found exposed on the ocean floor in transform faults, or along very slowly spreading ocean ridges. Sub-continental mantle can also be found exposed in some orogenic belts, and as xenoliths in some alkali basalts and kimberlites. While most mantle rock can be thought of as metamorphic, metamorphism can also occur during and after emplacement in the crust.

In addition to the mantle, ultramafic rocks can form in other ways, such as igneous cumulates (e.g., dunite, pyroxenite, hornblendite), metamorphism of some basaltic rocks, and extraction of felsic melt from a rock to leave behind an ultramafic restite. Metamorphosed ultramafic rocks can range widely in their hydration state from very dry (e.g., Fig. 12.1; a typical initial condition of fresh mantle and cumulate rocks from mafic magmas) to hydrated (e.g., Fig. 12.2).

The photomicrographs show an example of fresh olivine-rich ultramafic rock (Figs. 12.3A, B), a similar rock partially serpentinized at relatively low temperature (Figs. 12.3C, D), and one metamorphosed to amphibolite facies conditions (Figs. 12.3E, F). The last photomicrograph pair (Figs. 12.3G, H) is of a basalt protolith that was metamorphosed to become an ultramafic rock. The field photos show a similar organization. Figures 12.4–12.8 show fresh, mantle-derived ultramafic rocks. Figures 12.9–12.12 show similar rocks that have been serpentinized. Figures 12.13–12.16 show ultramafic rocks with different origins that have been metamorphosed at amphibolite facies conditions. Because of their diverse protoliths, chemical compositions, and metamorphic history, metamorphosed ultramafic rocks have similarly diverse mineralogy and appearance. The illustrations here give some characteristic examples.

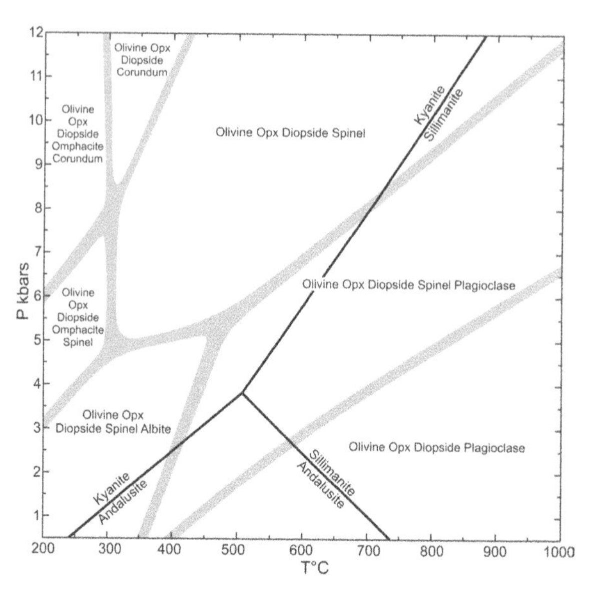

Figure 12.1 Phase diagram calculated for a dry ultramafic rock composition (here dry means no fluid phase), showing the fields of important assemblages. This can be thought of as a metamorphic grade-type diagram, like Figures 1.1 and 1.2, but not of such wide usefulness because of the large composition range of ultramafic rocks generally. The most important point is that aluminum is preferentially partitioned into low-density plagioclase at high temperature and low pressure, and into spinel at higher pressure and lower temperature. The digram was calculated for a depleted mantle composition: SiO_2, 44.90; Al_2O_3, 4.64; Fe_2O_3, 0.46, FeO, 7.83; MgO, 38.38, CaO, 3.50; Na_2O, 0.29, in weight %, modified after Salters and Stracke (2004). The diagram was calculated using Perple-X (Connolly, 2009). The aluminosilicate reaction lines are shown for reference purposes. Abbreviation: Opx, orthopyroxene.

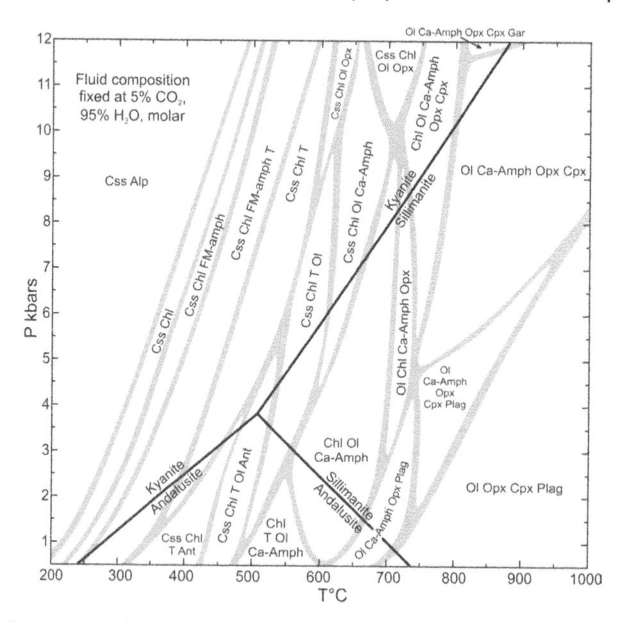

Figure 12.2 Phase diagram calculated for an ultramafic rock exactly like that in Figure 12.1, except that it was in equilibrium with a saturated fluid phase (5% CO_2 and 95% H_2O, molar %). The reason this figure has more fields than 12.1 is the two additional fluid components, CO_2 and H_2O. The more chemical components there are, the more possible phases there can be, so there are more possible combinations of phases (more fields). Like Figure 12.1, this diagram shouldn't be used to estimate metamorphic conditions for ultramafic rocks much different than this depleted mantle composition. There are some

Figure 12.2 (Continued) generalizations that are instructive, however. With increasing temperature and decreasing pressure, phases like chlorite and antigorite are progressively replaced by less hydrous phases like amphibole, and anhydrous phases like pyroxenes and olivine. Carbonate is also absent from the assemblages at higher temperature and lower pressure. Abbreviations: Alp, aluminous phase (e.g., pyrophyllite, kyanite), Ant, antigorite; Ca-amph, calcic amphibole; Chl, chlorite; Cpx, clinopyroxene, diopside or augite; Css, carbonate solid solution (Ca-Fe-Mg); FM-amph, ferromagnesian amphibole (anthophyllite or cummingtonite); Gar, garnet; Ol, olivine; Opx, orthopyroxene; Plag, plagioclase; T, talc.

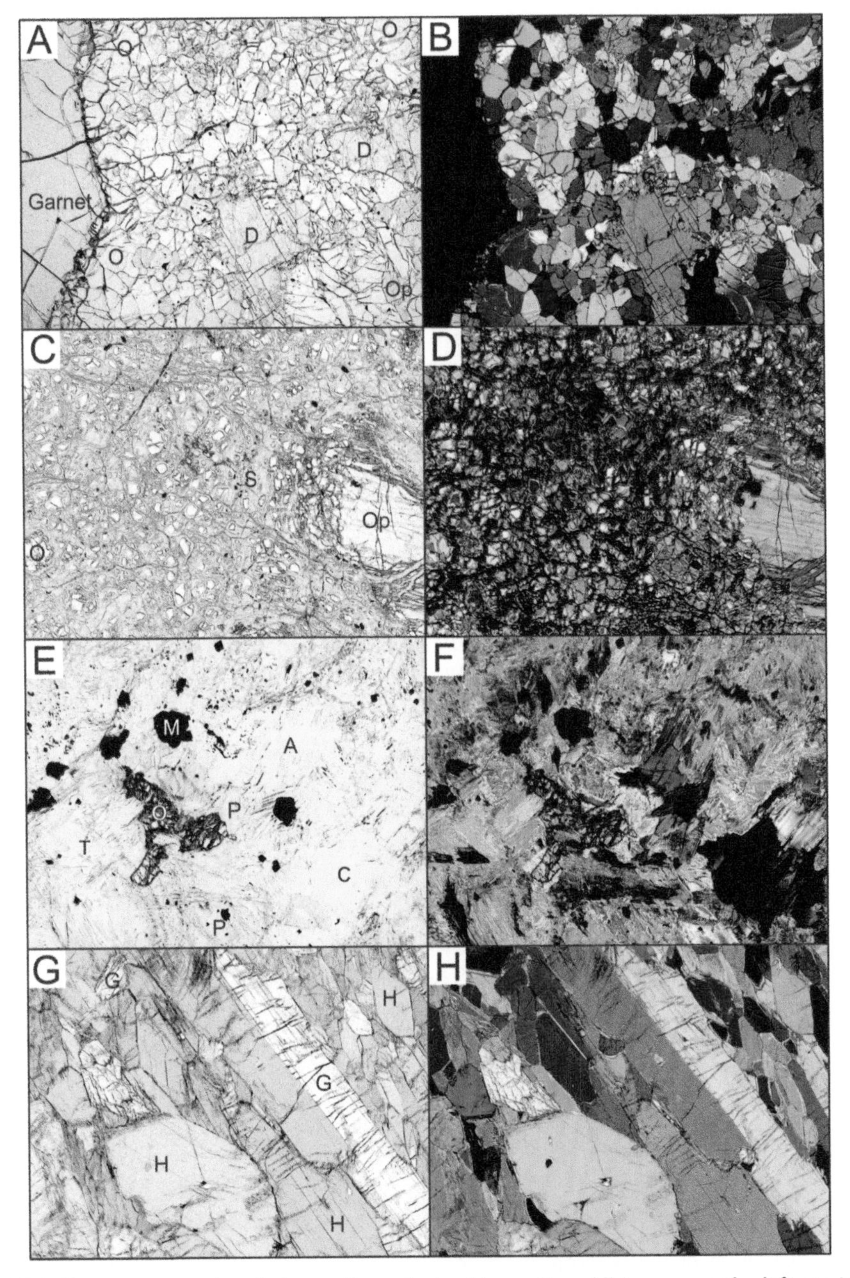

Figure 12.3 Photomicrographs of ultramafic rocks in thin section. All images on the left are in plane polarized light, and those on the right are the same fields in cross-polarized light. All field widths are 4 mm. A, B) Garnet peridotite in which the garnet is fresh, with only a thin dark reaction rim on garnet that indicates the start of breakdown of pyrope-rich garnet at low pressure. Note the pale-green chrome-diopside crystals. Almklovdalen, Åheim, Møre og Romsdal, Norway. C, D) Harzbergite from

Figure 12.3 (Continued) sub-oceanic mantle that has been hydrothermally altered, probably in the oceanic setting. Pale-green serpentine veins surround remnant grains of olivine and orthopyroxene. Bay of Islands ophiolite, western Newfoundland, Canada. E, F) Harzbergite that has been regionally metamorphosed to kyanite grade. Retrograde metamorphism after most deformation had ended probably formed this unfoliated, hydrous assemblage dominated by antigorite, talc, and chlorite. Pelham, Massachusetts, USA. E, F) Hornblende – gedrite rock derived from amphibolite facies metamorphism of an Mg-rich basalt, probably originally olivine-rich. Quabbin Reservoir, Massachusetts, USA. Abbreviations: A, antigorite; C, chlorite; D, chrome diopside; G, gedrite; H, hornblende; M, magnetite; O, olivine; Op, orthopyroxene; P, phlogopite; S, serpentine; T, talc.

Figure 12.4 Folded, layered mantle rock that was brought onto continental crust along a thrust fault. The brown-weathering rock is peridotite containing olivine, garnet, chrome diopside, and orthopyroxene. The layers within the peridotite are garnet pyroxenite, of approximately basaltic composition. Figure 12.5 is a close-up of the fresh garnet peridotite. Almklovdalen, Åheim, Møre og Romsdal, Norway.

Figure 12.5 Close-up of the outcrop in Figure 12.4, showing fresh garnet peridotite with a layer particularly rich in burgundy garnets and bright green chrome diopside. The fast-weathering olivine and orthopyroxene quickly develop a brown, rusty coating, so olivine does not have its classic olive green color and glassy luster. Almklovdalen, Åheim, Møre og Romsdal, Norway.

Figure 12.6 Close-up of the outcrop in Figure 12.4, in an area that originally had an assemblage like that in Figure 12.5. Here the garnet peridotite has been partially hydrated. Under crustal metamorphic conditions, the garnets partially reacted with infiltrating H_2O-rich fluid to make chlorite. Relict garnet cores can be seen in several grains, inside thick chlorite rims. The olivine is not significantly serpentinized because the chlorite grew under conditions outside of the serpentine stability field. Almklovdalen, Åheim, Møre og Romsdal, Norway.

Figure 12.7 Folded, layered garnet peridotite. The layers have differing proportions of olivine, pyroxenes, and garnet. Notice that, although the rock is covered with limonite stain, it is nearly free of lichen. This is a result of the rock being deficient in important nutrient elements such as phosphorous and potassium, and relatively rich in potentially toxic elements such as chromium and nickel. Ugelvik, Otrøy, Møre og Romsdal, Norway.

Figure 12.8 Many ultramafic rocks are distinctive on the landscape. Their low nutrient content and potentially toxic chromium and nickel tend to discourage lichens and other plants. Olivine- and orthopyroxene-rich ultramafic rocks stand out as brown rocks that starkly contrast with gray, lichen-covered rocks on the other side of the contact. Ugelvik, Otrøy, Møre og Romsdal, Norway.

Figure 12.9 Partially serpentinized dunite block in an accretionary wedge mélange. Numerous fractures that crosscut the dunite have been filled with fibrous serpentine (chrysotile). Serpentinite veins are also found at grain-scales, where grain boundaries are serpentinized, leaving intact olivine in some grain cores (like Figs. 12.3C, D). It is the green color of serpentine, and associated dark minerals like magnetite, that give this rock its color, rather than olivine. Sonoma County coast, California, USA.

Figure 12.10 Close-up of a serpentine vein in another part of the same block as in Figure 12.9, showing vein-crossing chrysotile fibers. The fibers grew in the direction that the vein opened, and layering within the vein indicates variable rates and directions of vein widening, and possibly changing fluid composition. The blue line shows the vein central surface, along which fluid once flowed, and the red line is parallel to the serpentine fibers, and so parallel to the opening directions. Sonoma County coast, California, USA.

Figure 12.11 Slickensided surface in serpentinite. The high ductility of serpentinite compared to most other rocks means that, during deformation, a lot of strain will be concentrated within and near the margins of serpentinite bodies. Weak shear planes commonly develop, producing the slickensides. While these can certainly be considered fault surfaces, they commonly extend for only a few centimeters or meters and so would not show up on any but outcrop-scale geologic maps. Sonoma County coast, California, USA.

Figure 12.12 Serpentine – talc-matrix breccia. Much like marble, weak, ductile serpentinite deforms easily and can take up a large amount of strain during deformation. Disruption of internal layers, and incorporation of wall rock fragments, can produce a tectonic breccia. Here there are brown-weathering masses and fragments of magnesite- or dolomite-rich blocks, rounded, black magnetite-rich serpentinite, and one elongate clinopyroxene-rich block just to the left of center. The latter has a carbonate-rich reaction rim around it. Warning! Serpentinites can be very slippery, so take great care while walking on them. Handöl, Jämtland, Sweden.

Figure 12.13 Part of a reaction zone between a large boudin of talc – magnesite ultramafic rock and garnet – muscovite schist host rock. The layering sequence in the transition zone, from the contact with schist inward, is biotite, biotite + actinolite, actinolite, actinolite + talc, talc, and talc + magnesite, with chlorite occurring throughout. Here, the visible minerals are biotite and some chlorite, plus crystals and radiating clusters of green actinolite in the center and bottom right. Chester, Vermont, USA.

Figure 12.14 Close-up of an olivine websterite boudin several meters long that was metamorphosed to upper amphibolite facies conditions, then partially retrograded at lower greenschist facies during the influx of H_2O-CO_2 fluid along fractures. This rock was originally made up of olivine and clinopyroxene, and is now mostly talc, magnesite, actinolite, chlorite, and antigorite. The large dark-greenish-gray crystals in the center are actinolite. The original olivine crystals are now pinkish, equant masses of serpentine, magnesite, and small amounts of hematite dust that give them their pink color (red arrows point out some). The rock protolith and high-grade metamorphism were Precambrian, but the small veins that crosscut the large actinolite crystals, and possibly the altered olivine crystals, are thought to be Ordovician. The age of the intermediate-grade actinolite – talc – chlorite – antigorite – olivine metamorphic assemblage is not very well constrained. Comstock, eastern Adirondacks, New York, USA.

Figure 12.15 Hornblendite boudins in hornblende-rich tonalitic gneiss. Although rocks like this can result from metamorphism of some basalt compositions (Figs. 12.3G, H), the chemical composition of

Figure 12.15 (Continued) this hornblendite is not typical of basalt. Instead its composition is consistent with it being a mafic cumulate (low Al, P, high Mg, Fe, Cr, Ni). It may originally have been a hornblendite cumulate, or a cumulate of anhydrous minerals that, on later hydration, became hornblendite. Kopparen, Fosen Peninsula, Sør Trøndelag, Norway.

Figure 12.16 Diopside-rich boudin made of approximately equal amounts of green clinopyroxene and black hornblende, in amphibolite facies quartzite. Quartzite is a metamorphosed sedimentary rock, so ultramafic rocks are not generally expected to be found in a sequence of them. Some basaltic rocks of the right composition, as dikes or lavas, can be metamorphosed to form ultramafic rocks (Figs. 12.3G, H), but probably not to clinopyroxene-rich rocks like this one. Indeed, this image shows two layers of typical plagioclase – hornblende amphibolite (yellow arrows), that are metamorphosed and highly deformed basalt dikes. Clearly these amphibolites do not have the right composition to be ultramafic metamorphic rocks, so an alternative explanation is necessary. Impure carbonate rocks are common in sedimentary successions. During metamorphism, decarbonation reactions can transform them to calc-silicate rocks. In this case, the clinopyroxene-rich boudin probably has a calcareous sedimentary rock origin. Interpreting the origins of ultramafic rocks should account for both their mineralogy and their geologic associations. Hasselvika, outer Trondheimsfjord, Sør Trøndelag, Norway.

Chapter 13

Contact metamorphic rocks

Contact metamorphism typically occurs where sedimentary or low-grade metamorphic rocks have been intruded by plutons, but they can also be found where such protoliths were brought into fault contact with hot crustal or mantle slabs. Heating of the fine-grained, reactive rock causes changes to rock mineralogy and texture, forming a region of metamorphic effects next to the heat source: the contact metamorphic zone or aureole.

The most common rocks to show the effects of contact metamorphism are shales, impure limestones, soils (such as under lava flows), and their low-grade metamorphic equivalents. Because affected rocks usually do not undergo much deformation during contact metamorphism, the resulting metamorphic textures tend to be isotropic, without preferred mineral orientations. Granofels is a typical type of contact metamorphic rock, referring to a granular rock with little or no preferred mineral orientation in which individual mineral grains are recognizable. Very fine-grained, flint-like contact metamorphic rocks are commonly called hornfels.

Figure 13.1 shows a mapped example of a contact metamorphic aureole surrounding a quartz monzonite pluton. The pluton is 6 to 8 km across, and intrudes regionally metamorphosed chlorite-grade rocks that include slate and phyllite. The contact metamorphic aureole ranges in width from 300 to about 1500 m. From the distant chlorite grade regional metamorphic assemblage, the aureole has biotite, andalusite, and sillimanite zones in sequence toward the pluton. The zones are defined by the first appearance of each indicator mineral in pelitic rocks.

The width of a contact metamorphic aureole depends on a number of factors, particularly the original composition and grain size of the rock prior to contact metamorphism, how hot the heat source initially was, the extent of the heat source, and the dip of the contact between the metamorphic rock and the heat source. If we consider only reactive rocks (e.g., shale, slate, impure limestone), a heat source at high temperature and large in extent, such as a pluton several kilometers across, may produce a contact metamorphic aureole hundreds of meters wide. In contrast, where the heat source was more limited, such as a small dike or sill, or the base of a lava flow, the metamorphic effects may be limited to slight induration over a few millimeters. In contrast, less reactive rocks such as quartzite or granite may not change substantially, even over many years at high temperature.

Figure 13.2A shows schematically the thermal changes in rocks originally having low ambient temperatures, following intrusion of a pluton. Initially, the temperature variation across the pluton is very sharp. Over time, heat diffusion cools the pluton and warms the surrounding rock. Over long periods of time, the rock temperatures decline to approach the original ambient temperature, unless there are other nearby heat sources in the interim. Figure 13.2B shows how three rocks at different distances

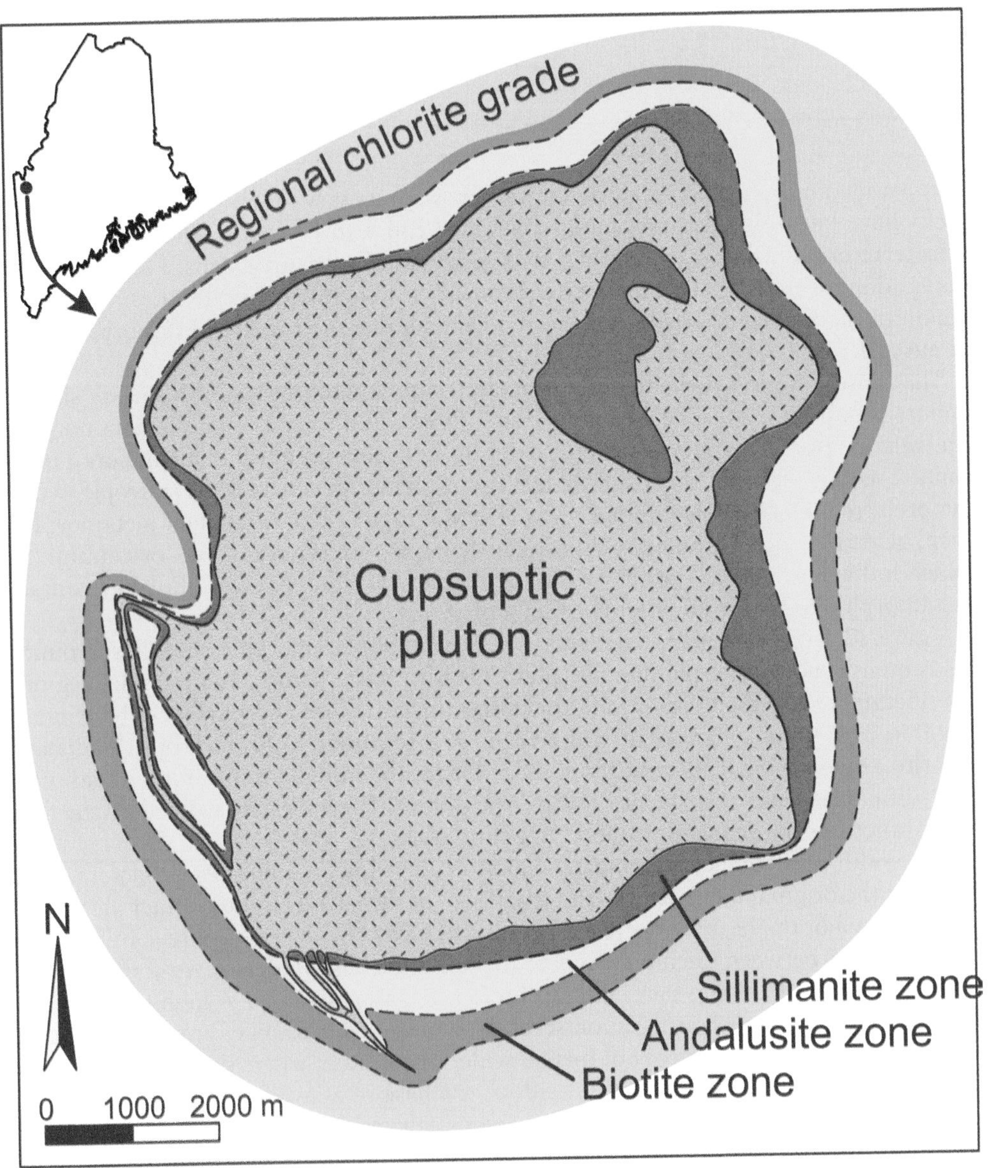

Figure 13.1 Simplified geologic and metamorphic map of the Cupsuptic pluton, western Maine, USA (after Harwood and Larson, 1969). This quartz monzonite pluton intruded regionally metamorphosed chlorite grade slate, phyllite, and other rocks. The contact aureole has biotite, andalusite, and sillimanite contact metamorphic zones, in sequence, inward toward the pluton from the chlorite grade rocks on the outside. Each zone is named after the indicator minerals, biotite, andalusite, or sillimanite, at their first appearance in pelitic rocks.

from the pluton change temperature over time. Rocks that don't get hot enough, for a long enough period of time, may have no noticeable contact metamorphic effects.

Figure 13.3 shows thin section images of two contact metamorphosed shales (Figs. 13.3A-D) and one basalt (Fig. 13E, F) that experienced different maximum temperatures. The field photos show contact metamorphosed pelitic rocks (Figs. 13.4–13.9), gneiss (Fig. 13.10), and basalt (Fig. 13.11). Figures 13.4–13.10 share a common feature: the contact metamorphic fabrics are superimposed on pre-existing layering.

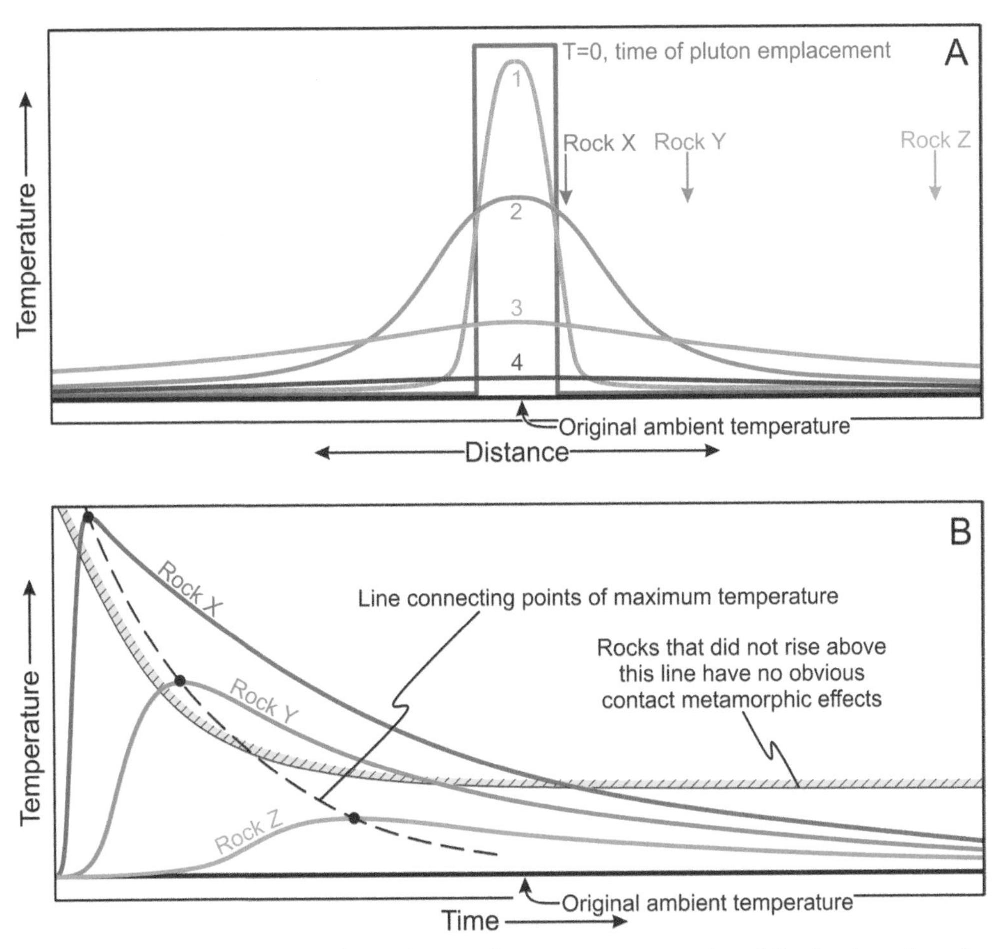

Figure 13.2 Schematic showing the evolution of temperatures over time following intrusion of a hot pluton into cooler rocks. A) At the time of pluton emplacement (T=0) there has been no significant heat loss to the surroundings. Over time, diffusion of heat progressively cools the pluton and heats the surrounding rock (T=1, 2, 3). After a long period of time, diffusion of heat to great distances brings the temperature of all rocks close to the original ambient temperature (T=4). B) Temperatures that rocks X, Y, and Z (Fig. 13.2A), at different distances from the pluton, experience over time. Contact metamorphic heating may produce no visible effects if temperatures were not high enough for a long enough period of time. The temperature curve for rock Z never rose above the yellow, striped line, and so will have no visible contact metamorphic effects. This schematic ignores the effects heat transfer by moving fluids, heats of reaction, etc. (see, for example, Suchý et al., 2004).

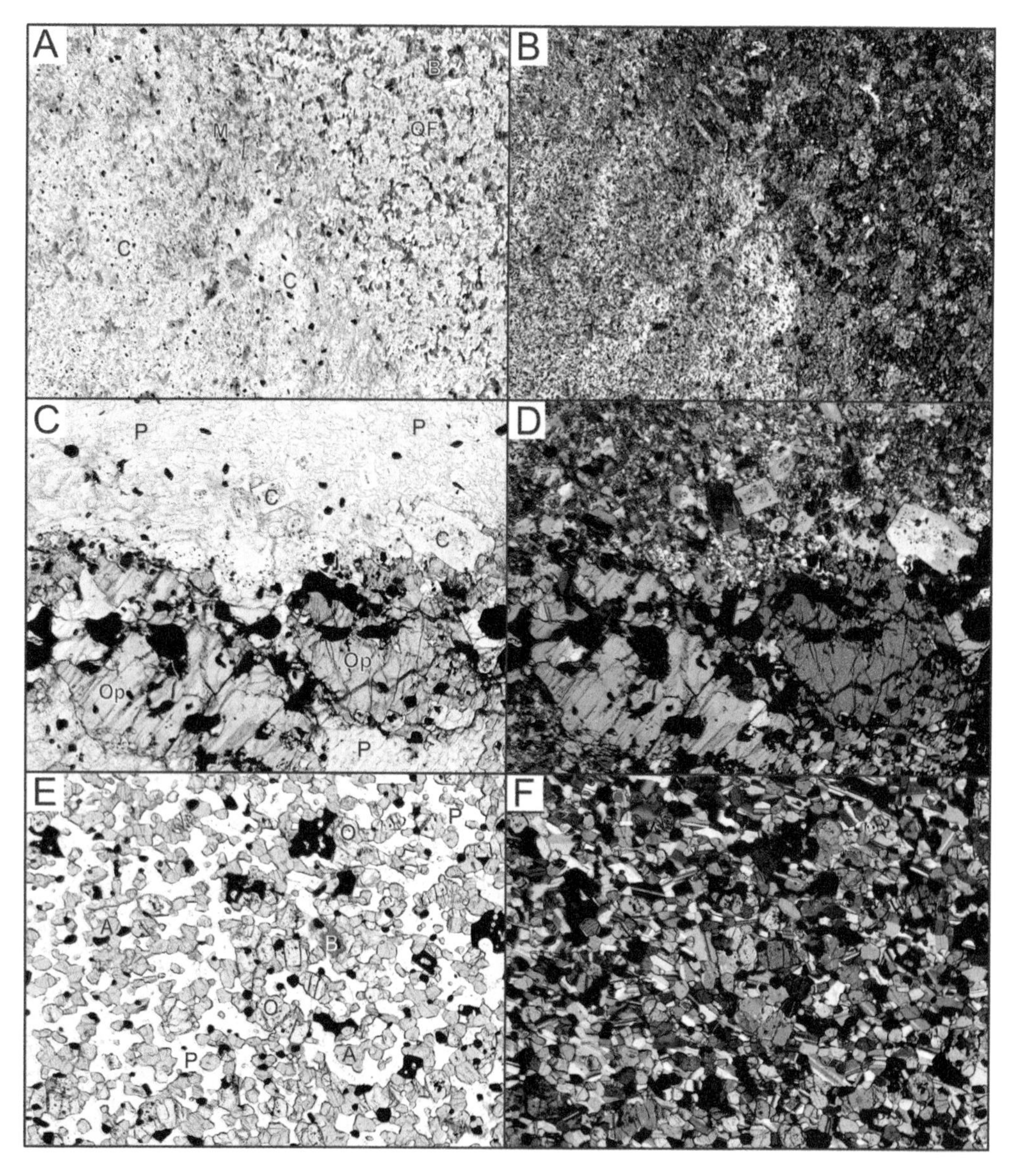

Figure 13.3 Contact metamorphic rocks in thin section. Images on the left are in plane-polarized light, those on the right are the same fields in cross-polarized light. All field widths are 4 mm. A, B) Contact metamorphosed slate in the contact aureole of a large granitic pluton. The slate has been transformed to a granofels, with the assemblage cordierite – K-feldspar – biotite – quartz – andalusite – plagioclase. Cordierite occurs almost entirely in large polycrystalline patches that are full of quartz, biotite, and feldspar inclusions (left half of the image). The largest grains in the matrix (right half of the image) are blocky, gray K-feldspar crystals. Andalusite also occurs, but not in this field. The small, plate-shaped crystals, colorless in A and colorful in B, are retrograde muscovite. Millinocket, Maine, USA. C, D) Contact metamorphosed slate near the margin of a gabbro pluton. This granofels is made almost entirely of blocky cordierite, fine-grained plagioclase, coarse, irregular orthopyroxene, and black magnetite. Quartz and K-feldspar components were probably lost as partial melts. York, Maine, USA. E, F) Contact metamorphosed basaltic lava, from a large xenolith block in a gabbro pluton. The original basaltic texture, with lath-shaped plagioclase crystals, has been recrystallized to form largely equant grains of plagioclase, augite, olivine, and magnetite. Skaergaard intrusion, east Greenland. Abbreviations: A, augite; B, biotite; C, cordierite; M, muscovite; O, olivine, Op, orthopyroxene; P, plagioclase; QF, quartz and feldspar.

Figure 13.4 Part of a large xenolith that underwent contact metamorphism in a granite pluton. The rock originally consisted of alternating layers of shale (layers with dark spots) and sandstone (fine-grained layers). The sandstone layers have been metamorphosed to biotite quartzite. The shale layers were metamorphosed to granofels, having the assemblage quartz – cordierite – andalusite – K-feldspar – biotite. During cooling, most of the cordierite and some K-feldspar reacted with H_2O fluid to form a fine-grained mixture of retrograde chlorite and muscovite, which weathers out as dark pits. Halifax, Nova Scotia, Canada.

Figure 13.5 Folded, layered phyllite, contact metamorphosed to granofels, with the assemblage quartz – cordierite – andalusite – biotite – orthoclase. Cordierite is the mineral weathering out as pits. Note that the rocks were folded during regional greenschist facies metamorphism, prior to contact metamorphism against a granite pluton. South Twin Lake, Maine, USA.

Figure 13.6 Prior to intrusion of a nearby gabbro pluton, this was a regionally metamorphosed slate at chlorite grade. During contact metamorphism, the foliated rock fabric was transformed to coarser, nearly isotropic granofels, and pink andalusite porphyroblasts grew as part of the contact metamorphic mineral assemblage. The andalusite crystals shown here are weathering out in raised relief on the rock surface. Greenville Junction, Maine, USA.

Figure 13.7 This photo has three different rock units: Dark gabbro above the yellow line, layered, contact metamorphosed Cretaceous sedimentary rocks between the yellow and blue lines, and Archean gneiss below the blue line. The sediments were probably folded during contact metamorphism, which must have approached magmatic temperatures. Note that the contact at the blue line is an unconformity. Skaergaard intrusion, east Greenland.

Figure 13.8 Close-up of one of the metamorphosed layers of Cretaceous sedimentary rock in Figure 13.7. The rock is mostly made of plagioclase, cordierite, and greenish spinel crystals. Skaergaard intrusion, east Greenland.

Figure 13.9 Andalusite crystals in fine-grained granofels in the contact aureole of the Bushveld Intrusion. Most of the andalusite crystals appear to be randomly oriented in the same plane as the flat rock surface, rather than being randomly oriented in three dimensions as might be expected. It may be that the andalusite crystals nucleated and grew from preexisting minerals that were part of an earlier foliation. South Africa.

Figure 13.10 Intrusion of a gabbro pluton caused melting of adjacent Archean gneiss for a distance of up to tens of meters. Movement of the partially melted gneiss broke up these refractory layers of amphibolite, injecting melt between the broken fragments. Skaergaard intrusion, east Greenland.

Figure 13.11 Close-up view of a basalt xenolith block that was contact metamorphosed in a gabbro pluton. The texture has been transformed to a granofels, with equant grains of plagioclase, augite, and olivine (like that in Fig. 13.3E, F, but a different rock). This is quite unlike typical basaltic textures that include lath-shaped plagioclase, interstitial and skeletal pyroxene, and fine-grained magnetite (some of these textures can survive metamorphism in some cases, see Figs. 9.16 and 25.17). Olivine in this rock tends to occur in polycrystalline clusters, possibly recrystallized phenocrysts, that weather out as pits. Skaergaard intrusion, east Greenland.

Chapter 14

Rocks in fault zones

Faulting involves localized, generally planar zones of high strain between large volumes of less-deformed or undeformed rock. At low pressures and typically low temperatures, rocks are brittle, and may be strong enough to sustain open fractures at least temporarily. At higher pressures and temperatures, rocks are more ductile, though ductility depends also on how fast the rocks were deformed. The slower a rock is deformed, the more likely it is to behave in a ductile way. Faults can have widths that range from less than 1 mm to over a kilometer. Wide faults are commonly referred to as fault zones, which may contain several discrete, narrower high-strain zones between less-deformed regions.

Brittle faults are characterized by sharp fracture surfaces, angular fragments within the fault zone (breccia, cataclastic texture), and commonly alteration of the fault zone as a result of open fractures allowing fluid access. Shear along the sharply defined fracture surfaces can result in extreme grain size reduction, forming microbreccias and even causing melting. Smooth, commonly shiny shear surfaces are called slickensides (Fig. 12.11). Slickenside surfaces typically have striations on them that parallel the fault slip direction.

In contrast, the higher pressures associated with ductile faults tends to inhibit open fractures unless fluid is available, at or near lithostatic pressure, to support them. Because they generally lack open fractures, ductile faults do not commonly have the hydrothermal alteration that is characteristic of many brittle fault zones. Despite that, ductile fault zones commonly have chemical compositions, mineralogy, or mineral proportions that differ from less deformed surrounding rocks. This difference may result from movement (addition and/or removal) of chemical components by flowing fluid. Alternatively, the ductile fault zone may have developed where it did because those rocks were originally different, and weaker, in the first place.

Ductile fault rocks are generally referred to as mylonites, defined as a ductile fault rock with reduced grain size compared to the parent rock. Protomylonite has less than 50% of the original grains reduced in size, mylonite (or mesomylonite) has 50–90%, and ultramylonite has greater than 90% (NAGM, 2004). The term 'mylonite' has no absolute grain size meaning except relative to the parental rock. It should be noted, however, that use of this term varies among field workers, and many include grain size as part of the working field definition. In such cases, 'ultramylonite' generally refers the most fine-grained ductile fault rocks. Mylonites as such may not survive to reach the surface if they remain hot for long periods of time, because recrystallization results in grain growth that can erase much or all of the mylonitic fabric.

Pseudotachylite is a very fine-grained, usually dark, flinty rock that typically occupies restricted parts of larger fault zones. The rock is essentially a microbreccia, with or without devitrified glass, on slip surfaces and filling extensional fractures. Pseudotachylites represent brittle fracture of a rock at very high strain rates, such as might occur during an earthquake. Pseudotachylite may enclose recognizable fragments of the adjacent rock.

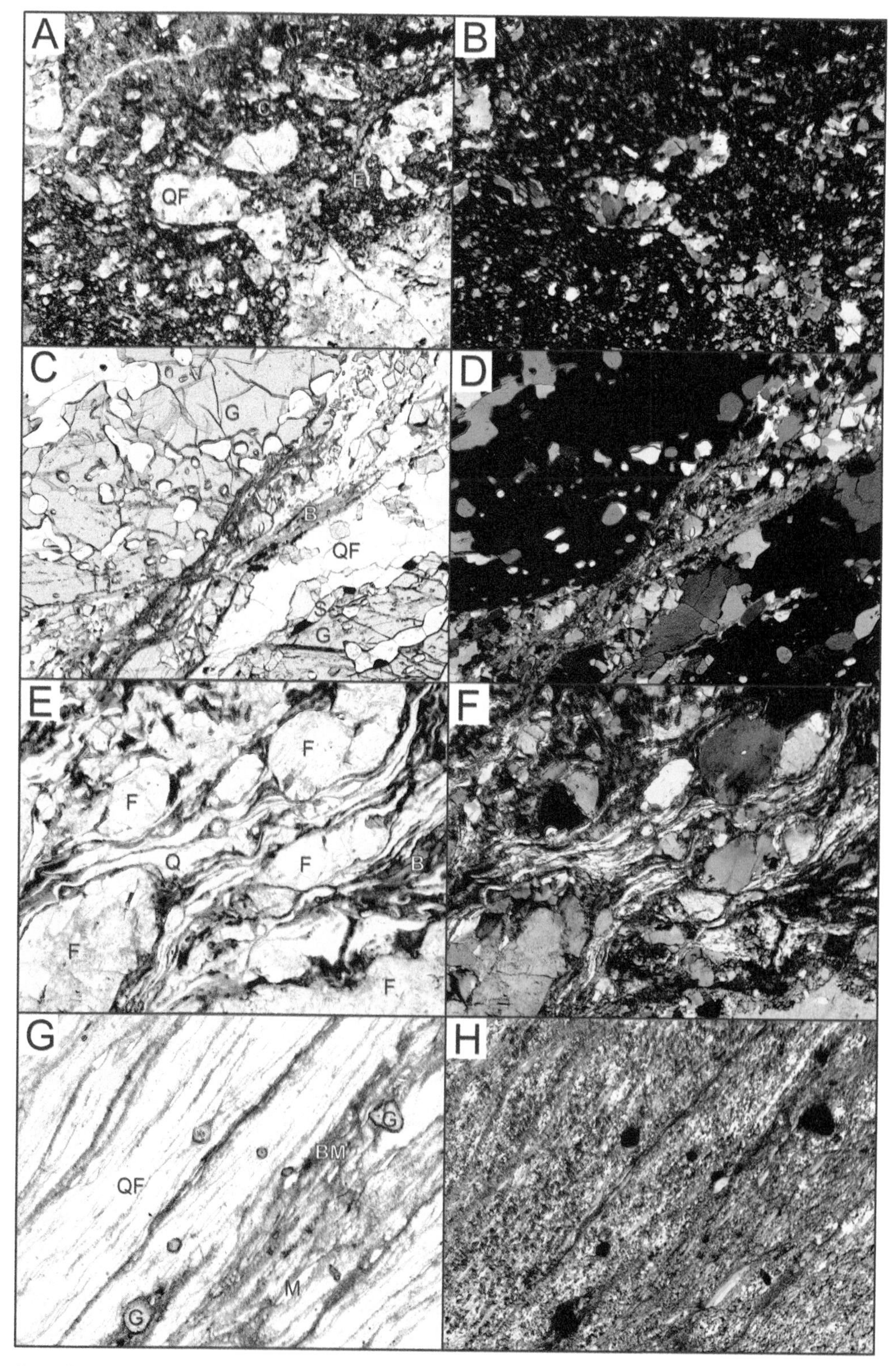

Figure 14.1 Fault rocks in thin section. All images on the left are in plane polarized light, those on the right are the same fields in cross polarized light. All field widths are 4 mm. A, B) Cataclastic fault rock derived

Figure 14.1 (Continued) from diorite. The characteristic features are angular fragments of all sizes, indicating brittle fracture. The mineral assemblage consists of quartz and feldspar clasts in a fine-grained matrix that contains, in addition, white mica, epidote, chlorite, and titanite. These minerals indicate greenschist facies metamorphic conditions during or after deformation. Saugus, Massachusetts, USA. C, D) Protomylonite with about 10% grain size-reduced rock, derived from a garnet – biotite – sillimanite – cordierite schist. Fine-grained shear zones cut across the center of the field from upper right to lower left. The shear zones have the same mineralogy as the rest of the rock, indicating that deformation took place at the same garnet – cordierite – sillimanite – biotite grade. Wales, Massachusetts, USA. E, F) Mylonite with about 70% grain size-reduced rock, derived from granodiorite. Thin ribbons of extremely fine-grained quartz and biotite weave between large porphyroclasts of feldspar. The mineral assemblage in the fine-grained matrix consists of quartz, plagioclase, microcline, biotite, muscovite, and epidote, indicating that deformation occurred in the epidote amphibolite facies. Marlborough, Massachusetts, USA. G, H) Ultramylonite (more than 90% grain size-reduced rock) derived from a muscovite – biotite – garnet schist. Most of the rock is has been drastically reduced in grain size, leaving small porphyroclasts of garnet and muscovite. Although the matrix grain size seems quite small, there has actually been some post-deformation recrystallization and grain growth. Many of the minute quartz crystals have approximately 120° grain boundary intersections. The quartz – muscovite – biotite – garnet mineral assemblage indicates that faulting took place at garnet grade. Brattvåg, western Moldefjord, Møre og Romsdal, Norway. Abbreviations: B, biotite; BM, biotite and muscovite; C, chlorite; F, feldspar; G, garnet; M, muscovite; Q, quartz; QF, quartz and feldspar; S, sillimanite.

The photomicrographs show one brittle fault rock (Figs. 14.1A, B) and three mylonites (Figs. 14.1C-H). The mylonites are in sequence, protomylonite, mylonite, ultramylonite, with progressively larger fractions of the rock having been reduced in grain size. The field photos first show examples of brittle faults (Figs. 14.2–14.4),

Figure 14.2 Brittle fault zone cutting coarse hornblende gabbro. The fault zone is made up of breccia which was infiltrated by fluid that deposited quartz and carbonate in the fractures, and hydrothermally altered thin adjacent zones of the host rock. The carbonate-bearing fracture filling and altered rock weathers to an orange color. White material is vein quartz that is not mixed with brown-weathering carbonate. The vertical lines are drill-hole traces. Hudson, Massachusetts, USA.

including pseudotachylites (Figs. 14.5, 14.6), followed by ductile fault examples (Figs. 14.7–14.16). Figure 14.17 is a cautionary example showing that faults are not always easy to spot. Many of the fault zones illustrated here are outcrop-scale and would not show up on a typical geologic map. Some people would call them local shear zones, or some other term, reserving the term 'fault' for map-scale features. This distinction is ignored here because it is the appearance that counts in this book, not the extent along strike.

Figure 14.3 This is a 10 cm wide brittle fault zone in granitic gneiss. The angular fragments have been altered along their rims to dark-green chlorite and brown-weathering dolomite, particularly visible on the flat, trapezoidal surface just to the left of the pen. Spaces between the breccia clasts are filled with brown-weathering dolomite and white quartz. Comstock, eastern Adirondacks, New York, USA.

Figure 14.4 A 1 cm wide brittle fault zone cutting black amphibolite and a white quartz vein, offsetting them by about 1.5 cm. The fault zone itself is made of several anastomosing fractures, some of which extend to the right of the main fault zone. The fault zone itself is made up of quartz, dolomite,

Figure 14.4 (Continued) chlorite, and pyrite. The amphibolite has been altered in 1 mm-thick zones adjacent to the fractures. This is one of a large number of small faults that are found near a map-scale fault that forms one side of a graben. The small faults have the same offset sense as the nearest map-scale fault. Paradox, eastern Adirondacks, New York.

Figure 14.5 Pseudotachylite (black) in a major detachment fault. The host rock is rusty-weathering mylonitic gneiss, with the pseudotachylite occurring in a layer approximately parallel to the mylonitic foliation (yellow line) and in crosscutting dikelets. The pseudotachylite probably represents very late fault movement, possibly unrelated to mylonite development. Hadlyme, Connecticut, USA.

Figure 14.6 Pseudotachylite crosscutting layering in a pyroxene-bearing granitic gneiss. Unlike igneous dikes, these pseudotachylite bodies are not laterally extensive, but instead commonly transition to thin mylonite or cataclastite zones at the ends. Nakkefjellet, Moldefjord, Møre og Romsdal, Norway.

Figure 14.7 Close-up of marble near that shown in Figure 4.6. The marble is the extremely fine-grained gray rock in the middle, with much coarser silicate rocks present to the left and upper right. The marble fabric is mylonitic, having been deformed at relatively low temperature as indicated by the lack of subsequent recrystallization and grain growth. This rock is much finer-grained than other marble, nearby, which escaped late deformation. Notice the small chips weathering out in raised relief on the marble surface. These are silicate rock fragments, either broken off the contacts with silicate rocks that flank the marble, or disrupted thin layers within the marble. Incorporation of silicate fragments into the mylonite has made it a small-scale tectonic breccia (compare to a larger-scale breccia in marble, Fig. 7.15). Bolsøy, Moldefjord, Møre og Romsdal, Norway.

Figure 14.8 A protomylonite shear zone (from top right to bottom left) in granodioritic gneiss. The gneiss contains mafic xenoliths (red arrows point out some) and both these and the gneissic fabric have been deformed by the shear zone. The shear sense is dextral, as indicated by the asymmetrically deformed mafic xenoliths at the shear zone edge, three of which are indicated by yellow arrows. The drill holes were for a magnetic anisotropy fabric study. Essex, Connecticut, USA.

Figure 14.9 Mylonitic fault rock derived from deformation of a coarse-grained granodiorite. All of the layers have approximately the same mineralogy: biotite, quartz, and feldspar. The darkest layers are mylonite to ultramylonite, whereas the lighter-colored layers have experienced less grain size reduction and are protomylonite to mylonite. Relict coarse, white microcline and plagioclase porphyroclasts have partially survived grain-size reduction during deformation. The red arrows point to fold hinges. Hudson, Massachusetts, USA.

Figure 14.10 Fault zone in granite, ranging from protomylonite to ultramylonite, containing numerous porphyroclasts of hornblende and feldspar. In general, the darker zones, such as that running vertically in the image center (red arrow), are finer-grained ultramylonite, in comparison to the lighter-colored zones that are protomylonite or mylonite. Dover, Newfoundland, Canada.

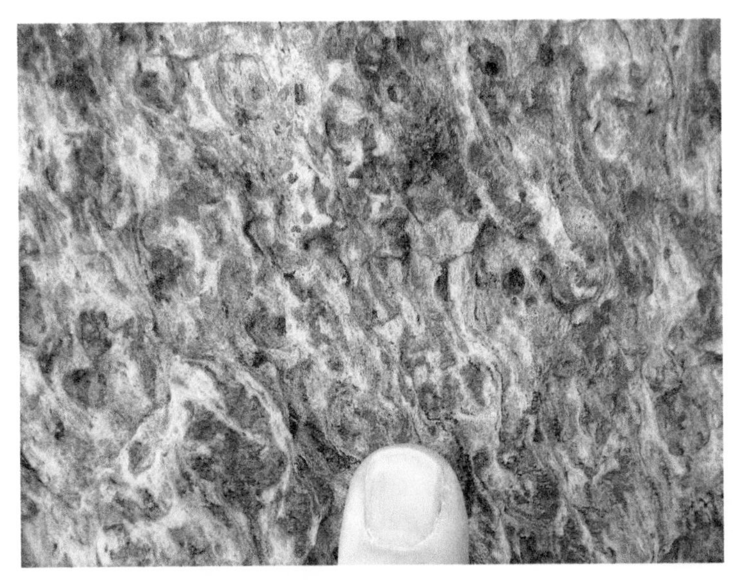

Figure 14.11 Mylonite in a gabbro that has been metamorphosed to granulite facies conditions, here shown perpendicular to the mylonite foliation. The fault zone has the same granulite facies metamorphic assemblage as the host rock. It is easy to see the reduced grain size matrix surrounding and wrapping around porphyroclasts of dark-green pyroxene and black magnetite. Haramsøy, Nordøyane, Møre og Romsdal, Norway.

Figure 14.12 This image is of the same mylonite as in Figure 14.11, but seen parallel to the mylonite foliation and lineation. The foliation surface has a strong vertical mineral lineation. Haramsøy, Nordøyane, Møre og Romsdal, Norway.

Figure 14.13 Mylonite in granodiorite gneiss, seen on a surface that is perpendicular to the foliation and parallel to the lineation. Asymmetry on small feldspar porphyroclasts (red arrows) indicates top-left shear sense. Quartz ribbons in the foliation plane are not visible in this view, but see the same rock in Figure 14.14. Høybakken, outer Trondheimsfjord, Sør Trøndelag, Norway.

Figure 14.14 Same mylonite as in Figure 14.13, but seen on a flat pavement surface parallel to the foliation and mineral lineation. Quartz ribbons are easily seen, smeared in the foliation plane, forming a lineation parallel to the red line. Høybakken, outer Trondheimsfjord, Sør Trøndelag, Norway.

Figure 14.15 Quartz diorite gneiss that has been transformed to mylonite in a zone about 30 cm thick (red arrow). The white material is vein quartz, with asymmetries on the vein quartz boudins that generally indicate a dextral shear sense. Kopparen, Fosen Peninsula, Sør Trøndelag, Norway.

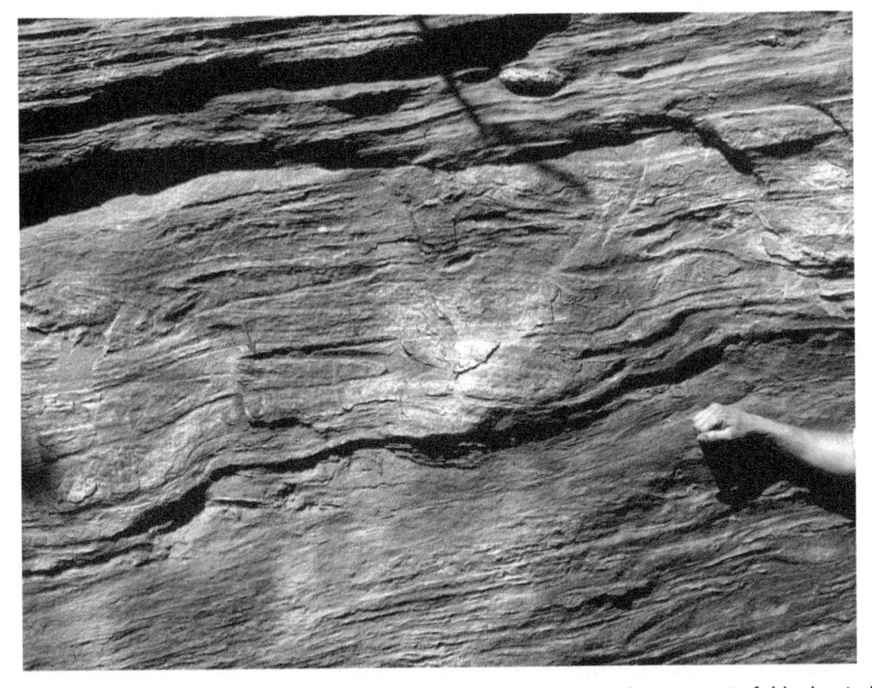

Figure 14.16 Section through a mylonitic fault zone that contains several asymmetric folds that indicate top-down-to-the-left shear sense. The most obvious fold hinge is identified with a red arrow, but there are many others. Above and below this zone is relatively straight foliation that contains small pegmatite boudins and isoclinal folds. Hadlyme, Connecticut, USA.

Figure 14.17 Frequently, faults are cryptic and must be inferred from interpretations of larger-scale geology. Here, the brownish-gray unit to the left is garnet – biotite schist of the amphibolite facies Blåhø Nappe, part of the Middle Allochthon of the central Scandinavian Caledonides. The greenish-gray unit to the right, on which the hammer is resting, is an epidote amphibolite facies amphibolite of the Støren Nappe, part of the Upper Allochthon. Based on a wide range of evidence from throughout the central Scandinavian Caledonides, these two units are interpreted to be the remnants of two different volcanic arc complexes, with the Støren originating farther outboard from Scandinavia than the Blåhø. The contact between the contrasting rocks and metamorphic grades is therefore a fault, along which there was considerable displacement. Though the rocks at the contact are certainly highly strained, they do not appear much different than rocks far from the contact. Bolsøy, Moldefjord, Møre og Romsdal, Norway.

Part 2: Metamorphic features

Chapter 15

Foliation, cleavage, lineation

The terms foliation and cleavage in metamorphic rocks have formal definitions, but their colloquial use in the field is commonly haphazard and interchangeable. Foliation refers to plane-parallel alignment of planar or plate-shaped features in a rock that developed during metamorphism. For minerals the word 'cleavage' refers to parallel planes of easy breaking. These are controlled by planes of particularly weak atomic bonds in the crystal lattice, spaced at an atomic scale. For rocks, cleavage in its most primitive sense refers to similar parallel-aligned planes or curved surfaces along which the rock breaks easily. Slate (Fig. 2.2) and flagstone (Fig. 2.7), are examples of rocks with good cleavage. In modern, colloquial use the term 'cleavage' has been expanded to include parallel sets of planar surfaces that developed during metamorphism, but along which rocks may not, in fact, easily break. That is, perhaps, the root of the foliation-cleavage terminology problem.

As shown in Figures 1.7 and 1.8, mineral foliation (parallel-planar alignment) generally develops by rotation of platy grains into parallelism during deformation. Compositional layering (a type of foliation, see Robertson, 1999) can develop several different ways and with different precursors, as indicated for gneisses in Figure 8.1, and for sedimentary rock protoliths in Figures 15.1A and B. Figure 3.5 is a quartzite with prominent compositional layering that may have originated as sedimentary bedding. However, deformation has strongly transposed the original bedding surfaces. The way this happens is that, during deformation, the original bedding surfaces (Fig. 15.1A) are offset along slip surfaces (transposed, Fig. 15.1B), so that the new apparent layering orientation (blue line) is different from the bedding surfaces (red lines). Figures 2.8 and 2.9 are examples of foliation defined by parallel alignment of plate-shaped minerals: muscovite, biotite, and kyanite in the first case, and muscovite and biotite in the second. Chlorite, talc, and antigorite are other examples of plate-shaped minerals that can define good foliations, but less platy minerals can too, such as chloritoid (Fig. 2.6, along with fine-grained chlorite and muscovite) and amphiboles (Fig. 10.5, along with phengite). Other features, like flattened cobbles (Fig. 7.4) and xenoliths (Fig. 7.13) can also define foliations.

There are many other terms commonly used to describe planar features in metamorphic rocks. For example, the foliation and cleavage in coarse-grained schist is commonly referred to as schistosity (e.g., Figs. 2.8, 2.9). Similarly, foliation, cleavage, and layering in coarse-grained gneissic rocks are typically referred to as gneissosity or gneissic layering (e.g., Figs. 8.4, 8.13). Here we will avoid these ancillary terms and try to use foliation to describe parallel alignment, cleavage to describe planes of easy breaking, and layering to describe a coarse foliation defined by mineralogical or compositional differences. Unsurprisingly, cleavages (easy breaking) are usually parallel to

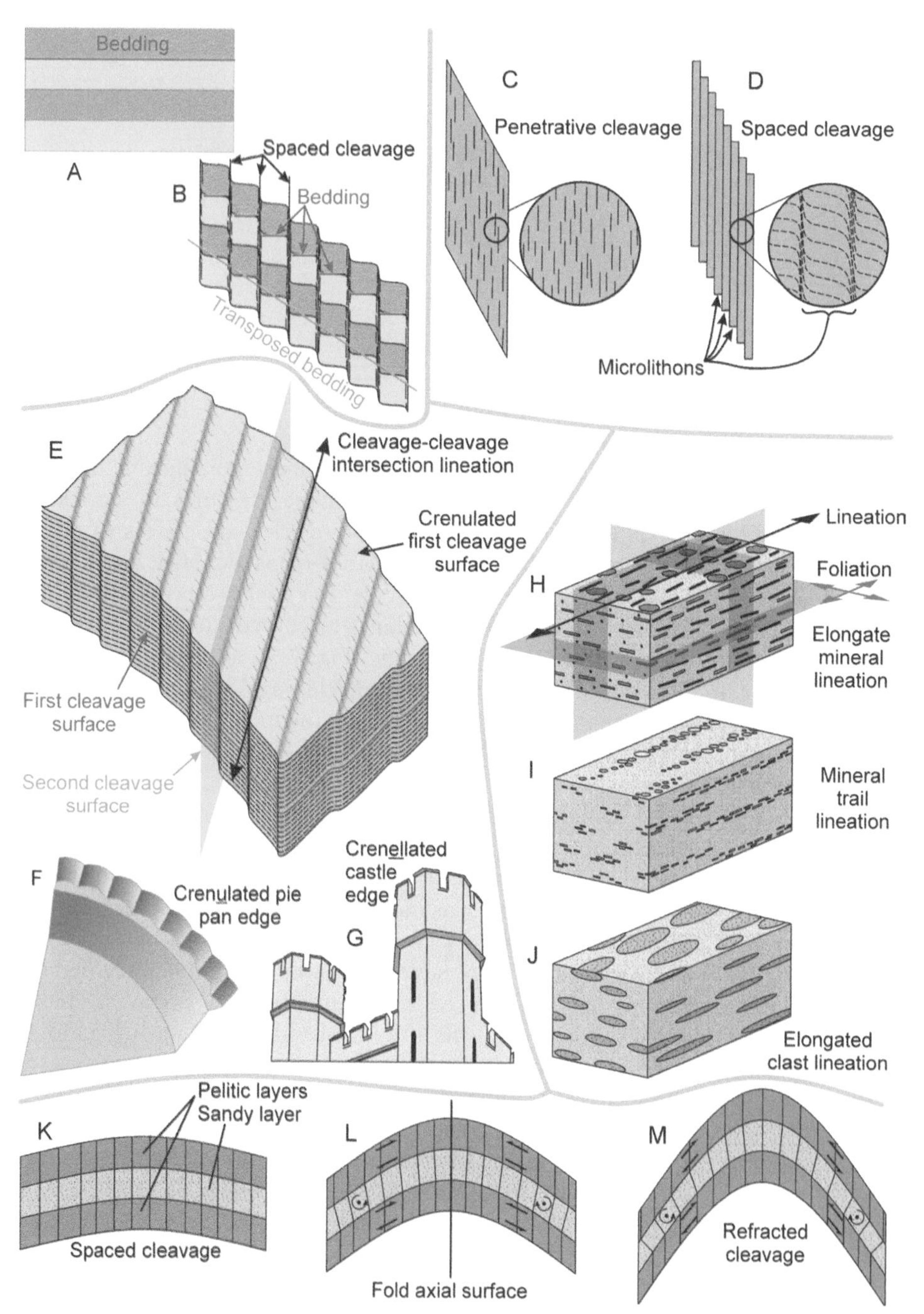

Figure 15.1 Schematic examples of cleavage and lineation. A) Bedding in a sequence of layered sedimentary or volcanic rocks, that are then metamorphosed and deformed by a spaced cleavage. B) The translation (movement) surfaces cut and offset bedding, producing transposed bedding that is at a different angle from actual bedding. Though technically incorrect, the term 'bedding' is commonly used anyway in the field. C) Closely spaced foliation elements (like oriented sheet silicates), that have no clearly separated translation planes, form a penetrative cleavage. Penetrative cleavages pervade the whole rock down to the scale of mineral grains. D) A spaced cleavage that contrasts with C in having translation planes that are clearly separated from one another, with relatively undeformed lithic elements in between (microlithons, like B). E) Crenulated foliation surface, caused by the intersection of an early, penetrative cleavage (red line) and a later spaced cleavage (blue surface). The early cleavage was bent by the later into little parallel folds

Figure 15.1 (Continued) (crenulations). The intersection produces a lineation, which is a set of parallel, line-shaped elements. This lineation is produced from the intersection of two cleavages, so it is a cleavage-cleavage intersection lineation. Note that crenulations are curved surfaces (F), as compared to crenellations (G), which are square notches. H, I, and J show lineations and foliations in the same orientation, but defined by different features. H has a mineral lineation defined by parallel, elongate minerals such as sillimanite or hornblende, in addition to a foliation defined by plate-like minerals such as biotite. In I, the lineation and foliation are defined by platy mineral fragments (e.g., muscovite) that are scattered along parallel trails. In J, pebbles or breccia fragments have been elongated and flattened parallel to one another, to produce both a lineation and a foliation. K, L, and M show how 'refracted cleavage' develops. Cleavage surfaces form (K) parallel to the fold axial plane early in fold development. As folding becomes more extreme, cleavage continues to develop and remain approximately parallel to the fold axial surface in the pelitic layers, where layer-parallel slip occurs. The cleavage in sandy layers becomes locked as the layers bend relative to the pelitic layers. The early cleavage in the more competent sandy layers therefore rotates with the fold limbs, so it forms an angle with the cleavage in pelitic layers.

foliations (parallel alignment). Where foliations are like pages in a book, cleavages are like the weak adherence of one page to another, and layers are like a stack of different books. For the field geologist, it is perhaps best to follow the descriptive path, where particular rock features like cleavages are described for what makes them up and how they look, rather than using just a single term and hoping you will remember later what it actually meant at that outcrop.

If cleavage planes occur throughout the rock, spaced on the scale of the mineral grains, it is referred to as a homogeneous or penetrative cleavage (Fig. 15.1C, and the first cleavage surface in E). If the planes are distinct and spaced more widely apart, they are referred to as a spaced cleavage (Fig. 15.1B, D, and the second cleavage surface in E), with spaced translation (movement, shear, slip) surfaces separating relatively undeformed rock segments called microlithons. A common occurrence during spaced cleavage development is that some mineral mass, most notably of quartz or carbonate, is lost from the shear surfaces. The lost material either escapes from the rock with metamorphic fluids, or is transferred to the microlithon centers. The newly developed composition difference between the shear surfaces and microlithons represents a new foliation feature (layering), and is a type of metamorphic differentiation.

Because rocks are weak in the cleavage orientations, and so tend to break along them, these planes can be exploited for economic purposes to produce roofing slate, flagstone, and even to help produce large blocks of weakly-foliated building stone. Cleavages can also be the planes along which landslides and fault slip can preferentially occur. Differential weathering along cleavages can result in landscapes that are partly controlled by cleavage orientation.

In contrast to foliation, cleavage, and layering, which are planar features, lineations are parallel sets of line-like features, rather like a handful of pencils or straw. To use the book analogy, lineations on a foliation surface are analogous to parallel sets of lines on book pages. There are two broad types of metamorphic lineations. The first type are intersection lineations, where planar surfaces of different orientations intersect along lines, such as the intersection of a newer cleavage set with an older set (Fig. 15.1E). The second lineation type is defined by oriented solid parts of the rock (e.g., minerals, clasts, xenoliths), such as the parallel alignment of elongate grains like hornblende or sillimanite (Fig. 15.1H), elongate clusters of grains like quartz ribbons or trails of recrystallized muscovite (Fig. 15.1I), or elongate clasts or xenoliths (Fig. 15.1J). Together, foliations, cleavages, and lineations record some of the deformation history of the rock, measurements of which are essential for structural interpretations.

The first two thin section images (Figs. 15.2A, B), show fine-grained pelitic rocks, each having a dominant (obvious) foliation and cleavage, and a second that is less obvious. Figures 15.2C and D show weakly foliated rocks, a hornblende amphibolite and a biotite gneiss, respectively, that have little or no rock cleavage. Figures 15.2E and F show rock lineations defined by elongate crystals of gedrite and sillimanite, respectively. In the field photos, Figures 15.3–15.6 show rocks with single visible

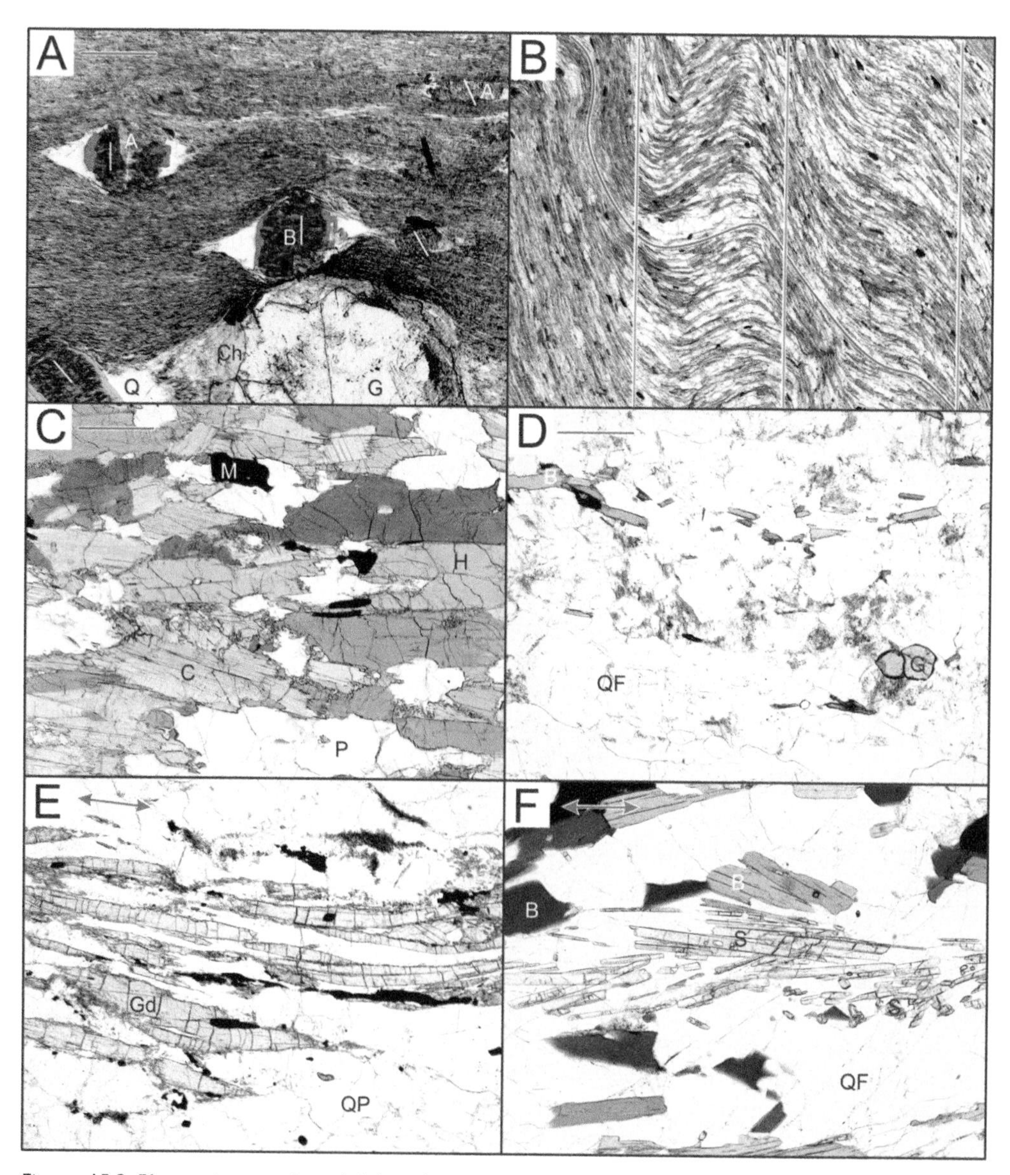

Figure 15.2 Photomicrographs of foliated and lineated rocks in thin section. All images are in plane-polarized light and have field widths of 4 mm. A) Fine-grained phyllite with an obvious, late, penetrative foliation and cleavage (red line) that is distorted somewhat around porphyroblasts

Figure 15.2 (Continued) of garnet, biotite, and albite. There is an earlier foliation, overgrown by biotite and albite (yellow lines), that is at a high angle to the obvious foliation. The earlier foliation is defined by graphite inclusion trails in biotite and albite, and by biotite cleavage orientations in porphyroblasts. Note the prominent quartz strain shadows on biotite, produced by vertical flattening and horizontal extension of the rock. There is also minor retrograde chlorite on the garnet. Charlemont, Massachusetts, USA. B) Muscovite-chlorite phyllite having a strong early penetrative foliation and cleavage (red line) that has been folded by a later spaced crenulation cleavage, parallel to the blue lines. New Salem, Massachusetts, USA. C) Hornblende – cummingtonite amphibolite, cut perpendicular to the weak foliation, showing aligned amphiboles. Because the alignment is not very good, and because amphibole has two moderately good mineral cleavages rather than a single perfect one like micas, the rock itself has almost no cleavage. Quabbin Reservoir, Massachusetts, USA. D) Granitic gneiss having a weak foliation defined by sparse biotite. Because of the scarcity of biotite, the rock itself has essentially no cleavage. Northfield, Massachusetts, USA. E) Amphibolite having a strong lineation defined by the parallel alignment of gedrite crystals, approximately parallel to the red arrow. Orange, Massachusetts, USA. F) Sillimanite lineation in a sillimanite – orthoclase – biotite – quartz schist. The lineation direction is approximately parallel to the red arrow. Manchester, New Hampshire, USA. Abbreviations: A, albite, B, biotite, C, cummingtonite; Ch, chlorite, G, garnet; H, hornblende; M, magnetite; P, plagioclase; Q, quartz; QF, quartz and feldspar, QP, quartz and plagioclase.

Figure 15.3 Weakly-developed spaced cleavage at lower greenschist facies, in pelitic layers that are sandwiched in between graded turbidite layers. In the upper part of the image the cleavage is best seen just to the left of, and parallel to, the solid red line. This area is expanded in the inset. There are also several fractures, dipping down to the right, beneath the overhang and parallel to the red line, that are manifestations of the cleavage (blue arrows). In the lower part of the image the cleavage is

Figure 15.3 (Continued) defined by widely spaced fractures parallel to the dashed red line, in the same orientation as the cleavage and fractures in the upper part of the image. The central, quartz-rich layer has quartz veins (yellow arrows) that have approximately the same orientation as cleavage in the pelitic layers. The layers have been folded but here are stratigraphically right-side-up. Yes, you may be justified calling these sedimentary rather than metamorphic rocks. Teveltunet, Nord Trøndelag, Norway.

cleavages or foliations. Figures 15.3 and 15.4 show weakly developed cleavages in slate and amphibolite, respectively. The weak nature in the first case is the result of relatively little deformation at low metamorphic grade. In the second case the weak foliation is partly caused by mineralogy, where coarse, blocky hornblende and plagioclase do not usually sustain good foliations or cleavages. Figure 15.5 shows strong cleavage development, but with cleavage refraction in different materials (illustrated in Figs. 15.1K-M). Figure 15.6 is of a granular rock that has no cleavage, but a strong foliation (layering). Figures 15.7–15.10 show rocks with two cleavages cutting one another: an early cleavage set that is cut and deformed by a later cleavage set. Figures 15.11–15.17 (and 15.7) show both cleavages and lineations: 15.11–15.13 have mineral lineations, 15.14–15.16 have cleavage-cleavage intersection lineations, and 15.14 and 15.17 have both.

Figure 15.4 Although platy minerals, such as the micas and chlorite, dominate metamorphic foliation and cleavage, they can be defined by other minerals as well. Here, an amphibolite has a weak horizontal foliation, parallel to the red line, defined by oriented hornblende crystals, that is axial planar to the folded pegmatite dike (blue form line shows the fold). The weakness of the foliation, and the lack of a single perfect cleavage in hornblende or associated plagioclase, means that this example has no notable rock cleavage. Warrensburg, Eastern Adirondacks, New York, USA.

Figure 15.5 Cleavage development and 'refraction' in metamorphosed pelitic and quartz-rich layers, as seen on a weathered joint surface of greenschist facies rocks. Cleavage is better developed in the fine-grained pelitic layer, spanning the middle of the image (yellow arrow) than in the quartz-rich layers above and below. As rock folding progressed, cleavage in the pelitic layer continued to develop and remained approximately parallel to the fold axial surface (parallel to the blue line). In the quartz-rich layers (top and bottom) the cleavage became locked and simply rotated with the fold limb. The result is that the cleavage trace (red line) on this joint surface is bent (refracted). Badger Head, Tasmania.

Figure 15.6 Foliated (or layered) gabbroic gneiss. The alternating layers are made up of fine-grained pyroxene (dark) and plagioclase (light). The original coarse grain size of the gabbro protolith is indicated by the relict large, dark pyroxene augen (red arrows points out two, see Chapter 17), suggesting that the layering resulted from ductile deformation of layered gabbro at high temperature. The fine-grained pyroxene and plagioclase crystals in this rock are more or less equant, and have two cleavages that are not particularly good. As a result, this rock only has a foliation (layering), with no cleavage. Giles Complex, central Australia.

Figure 15.7 Foliation and cleavage surface in muscovite – biotite schist, reflecting the sunlight. The foliation is the result of parallel alignment of biotite and muscovite crystals. The alignment and excellent mineral cleavages of muscovite and biotite allows the rock to break along the foliation, so the rock cleavage is parallel to the foliation. Some of the irregularities in the foliation are caused by local deformation around quartz vein boudins. Small folds on the shiny cleavage surface, surrounding and parallel to the red line, are a weak cleavage-cleavage intersection lineation. Big Thompson Canyon, Colorado, USA.

Figure 15.8 This rock has two cleavages that cross one another, seen on a joint surface. The left part of this photo is schist and the right part is quartzite, with the contact located between the yellow arrows. The schist has one early penetrative cleavage that runs from upper left steeply downward and to the right (yellow, wavy line), which intersects the schist-quartzite contact at a small angle (blue line). A later spaced cleavage (red line) cuts the earlier foliation and folds it to produce S-shaped crenulations (why the yellow line is wavy), an asymmetry indicating dextral slip (left side up and to the right). The quartzite to the right has the early cleavage weakly developed (dashed yellow line), visible as faint cracks approximately parallel to the schist-quartzite contact. The later spaced cleavage does not visibly penetrate far into the quartzite layer. The field width is about 15 cm. Windsor, Massachusetts, USA.

Figure 15.9 This granitic gneiss has a relatively strong early foliation (or gneissic layering) and a weak, parallel cleavage (yellow line). The early cleavage is cut at a high angle by a late spaced cleavage (red line). The spaced cleavage surfaces fold the older foliation. Figure 15.10 is a close-up of the region next to the lens cap. Big Thompson Canyon, Colorado, USA

Figure 15.10 This is a close-up of the early foliation (gneissic layering, weak cleavage, yellow line shows its general form) and the later spaced cleavage (red line) seen in Figure 15.9. The later cleavage surfaces are irregularly spaced 3–10 mm apart. The original foliation is folded along these cleavages, producing small crenulations (blue line). Big Thompson Canyon, Colorado, USA.

Figure 15.11 Tonalitic gneiss having a vertical cleavage (the flat wall parallel to the plane of the hammer head and handle) and strong vertical mineral lineation on the cleavage surface (red line). The rock contains mostly plagioclase and quartz, with only 10% biotite and other minerals. The strong lineation is made up of long ribbons of quartz about 1 mm wide and strings of recrystallized plagioclase and biotite. The rock surface below and to the right of the hammer point is approximately perpendicular to the foliation and lineation, so the lineation is seen almost end-on (just looks like black-and-white speckles) and the foliation is clearly weak. Almvikneset, Trondheimsfjord, Sør Trøndelag, Norway.

Figure 15.12 Lineated and foliated granitic augen gneiss. A strong lineation, parallel to the pen, is exposed on a joint surface that is close in orientation to the foliation (red line). The lineation and foliation are defined by extended and largely recrystallized pink and light-gray feldspar augen and the thoroughly recrystallized gray quartz – biotite matrix. The rock has almost no cleavage. Comstock, eastern Adirondacks, New York, USA.

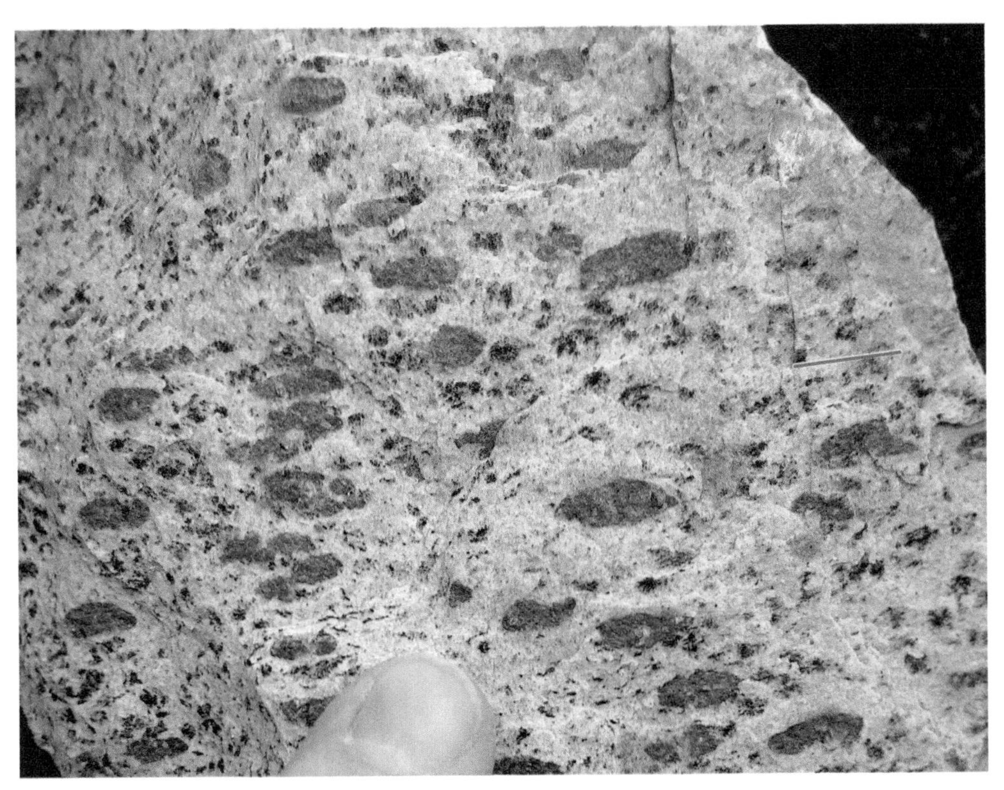

Figure 15.13 A somewhat weathered, fine-grained felsic gneiss, that has flattened and elongate, rusty-weathering pyrite grains. The pyrite grains define both the rock lineation, parallel to the red line, and, with minor biotite, the foliation. The foliation surface occupies the entire center of the image. Rissa, outer Trondheimsfjord, Sør Trøndelag, Norway.

Figure 15.14 A fine-grained amphibolite that has an early penetrative cleavage (flat rock face on which the hammer handle is resting) that has on it a prominent mineral lineation parallel to the red line. The early cleavage surface is cut by a later spaced cleavage (parallel to the yellow line), resulting in a strong cleavage-cleavage intersection lineation on the early cleavage surface (parallel to the yellow arrow). There is a later, irregularly spaced cleavage that is parallel to the blue line, which is at a slight angle to the yellow line cleavage. Bolsøy, Møre og Romsdal, Moldefjord, Norway.

Figure 15.15 Fine-grained pelitic schist showing a flat cleavage surface defined by a strong muscovite and biotite foliation. The flat cleavage surface is cut by a relatively coarse crenulation cleavage parallel to the red line, and a finer-scale crenulation cleavage parallel to the yellow line. Both form cleavage-cleavage intersection lineations on the early cleavage surface. Charlemont, Massachusetts, USA.

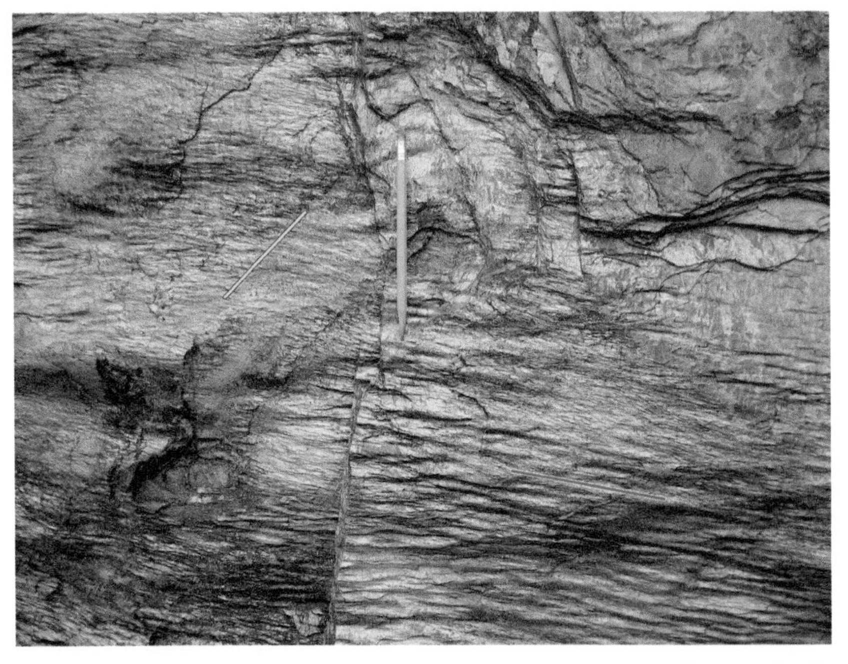

Figure 15.16 Folded phyllite at greenschist facies. The shiny early cleavage surface is defined by white mica and chlorite. It has been cut by a spaced crenulation cleavage set that intersects the early cleavage at a high angle, to produce small folds, crenulations and larger, parallel to the red line. Lithologic inhomogeneities are visible parallel to and to the right of the yellow line. These may be related to bedding in the protolith. Heggset, Sør Trøndelag, Norway.

Figure 15.17 Phyllite at chlorite grade that has a strong penetrative early cleavage. The cleavage is defined by white mica and chlorite grains that are almost too small to see, but their parallel alignment (foliation) produces satin-like cleavage surfaces. The red line shows the form of this early cleavage where it is exposed along a crosscutting break. There is a faint mineral lineation (blue lines) on the early cleavage surface that is defined by elongate trails of sheet silicates and oriented tourmaline crystals. The early cleavage is cut by a spaced cleavage that produces small folds (crenulations and larger), parallel to the yellow line on the early cleavage surface. The field width is about 15 cm. Petersburg, New York, USA.

Chapter 16

Folds

Folds are perhaps the most distinctive feature of deformed rocks. Folds occur in all kinds of rocks, including sedimentary, igneous, and metamorphic. Folds in sedimentary rocks are perhaps best known in mountain fold and thrust belts. In igneous rocks, the folded surfaces of pahoehoe lava flows are perhaps the best example, and folded schlieren (wispy dark layers) in many light-colored plutonic rocks are another. In metamorphic rocks folds can be defined by almost any curved surface, including foliation, compositional layering, cleavage, dikes, veins, or geologic unit contacts. Folds come in all sizes, ranging from delicate crenulations on a foliation surface (Figs. 15.15–15.17) to folds many kilometers across. They also come in a variety of shapes, from graceful symmetric arcs to confusing messes reminiscent of a failed origami project. Folds illustrated here are defined by curved layers, curved foliation surfaces, and even curved fold axial planes.

Folds have a wide range of three-dimensional shapes, and shape characteristics can be used to classify them and to extract structural information. One important characteristic of folds is the ratio of fold amplitude to wavelength. Low ratios define open folds (Figs. 16.1A–C, and D on the left side), and high ratios, at the extreme, describe isoclinal folds for which the two fold limbs are nearly parallel (Fig. 16.1D on the right side). Another important fold characteristic is symmetry. If the folded layers on one side of an axial surface are a mirror image of the other, the fold is symmetrical. If the two sides are different, the fold is asymmetrical. Asymmetry can indicate the local or regional shear sense during fold development (Figs. 16.1B, E, F). Similar folds, such as those shown schematically in Figures 16.1A–C have nested layers that don't change substantially from one layer to the next. Similar folds are typical of ductile deformation in general, but other types can be found in metamorphic rocks as well, including concentric folds (Fig. 16.1G, also 3.5), accordion-like ptygmatic folds (Fig. 16.1H), and sheath folds (Fig. 16.1I). Because folding requires rock strain (change in shape), folding is commonly associated with the development of axial planar foliation and cleavage (Figs. 16.1A, C).

Many regions have been subject to more than one episode of folding, and usually the stress orientations during the different fold episodes were also different. Multiple episodes of deformation result in fold interference and refolded folds. In these, the earlier axial surfaces can no longer be flat planes (Fig. 16.1D, right side), and the fold axes are unlikely to remain straight lines.

Figure 16.2 shows small folds visible in thin section, which can also be seen with a hand lens on broken or cut rock surfaces. These examples are of fine-grained rocks because such small-scale folds usually must be in rocks where the grain size is smaller than the folds themselves (but see Fig. 10.1A). The field photos are arranged to show similar, open folds first (Figs. 16.3–16.5), followed by similar folds with sharper

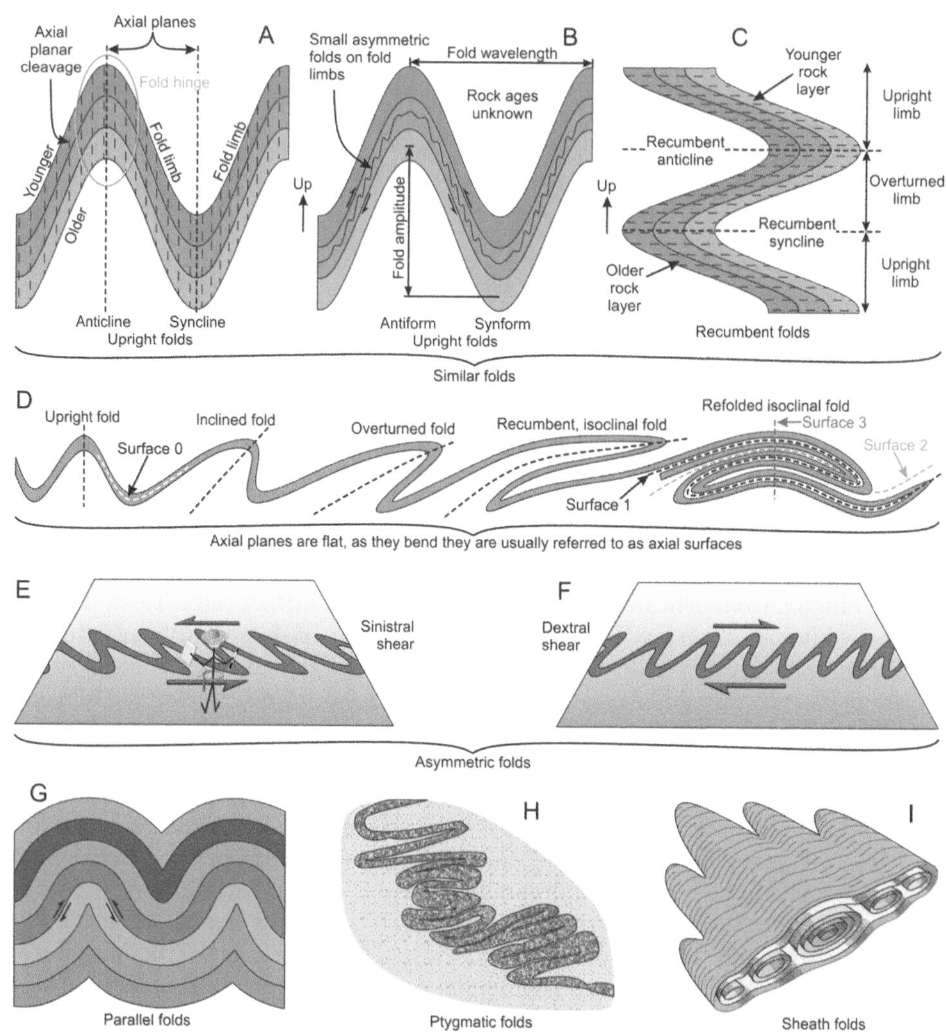

Figure 16.1 Fold anatomy and examples of fold types. A) Typical upright, symmetric, similar folds, showing layer ages, axial planes, axial planar cleavage, and the locations of fold hinges and limbs. Here the stratigraphic sequence is known and upright, making it unambiguous which folds are anticlines and synclines. B) The same folds as A, but showing the shear sense along fold limbs that can result in asymmetric minor folds. If layer ages are unknown, or if ages reverse or repeat in order because of folding and faulting, the words synform and antiform are used instead of syncline and anticline. C) Recumbent folds (lying on their sides), where the terms antiform and synform lose their meaning. If age relations are known, however, they can still be classified as recumbent synclines and anticlines. D) Progression of fold terminology with increasing deformation and overturning of a folded layer (e.g., a sedimentary rock bed, surface 0; axial surfaces are only shown for antiforms). On the far right the original recumbent, isoclinal fold is refolded along two additional axial surfaces in numbered sequence: the first fold is isoclinal (1, black, also present in all folds to the left), the next is also isoclinal (2, blue), and the last fold is open (3, red). E) Asymmetric folds with arrows indicating the local shear sense that produced the asymmetry. As is the custom with faults, the shear sense refers to the motion of the distant side as compared to the near side. In this case the shear sense is sinistral (left-lateral, far side moved to the left). F) The same as E, but showing a dextral sense of shear. G) Parallel folds that result from constant thickness layers that slip along shear surfaces between them (see the bottom of Fig. 3.5). H) Ptygmatic folds, that result from compressional shortening of a layer that was stronger than the enclosing rock, causing the strong layer to buckle and fold irregularly. I) Sheath (or tubular) folds that develop during extreme deformation.

fold hinges and fold limbs closer to being parallel to one another (Figs. 16.6–16.11). Following those are asymmetric folds (Figs. 16.11–16.14, 16.11 pulling double duty here) that indicate shear along the folded layer. Figures 16.15–16.18 show fold interference and refolded folds, in which an earlier set of folds and their axial surfaces were refolded by another set. Finally, examples of ptygmatic folds (16.19, 16.20) and sheath folds (Figs. 16.21, 16.22) are shown. The great variety of folds and fold characteristics makes it impossible to do justice to all of them, but this selection illustrates some representative examples.

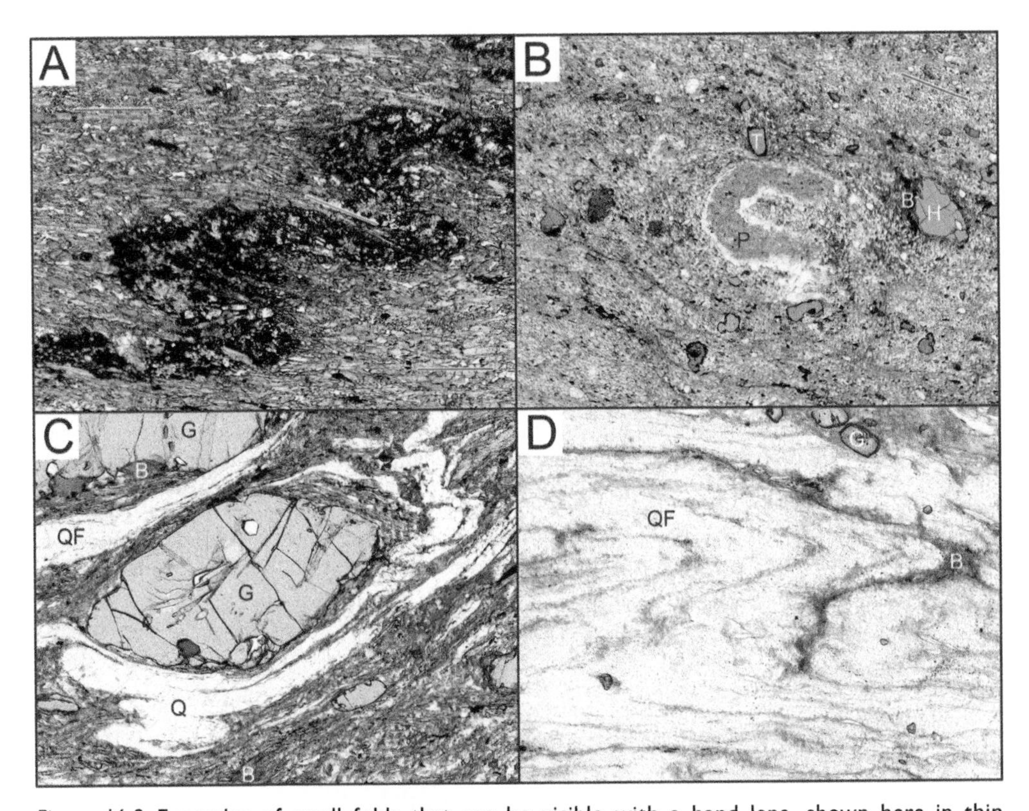

Figure 16.2 Examples of small folds that can be visible with a hand lens, shown here in thin section. All images are in plane-polarized light and have field widths of 4 mm. Thin sections were cut parallel to the rock lineation and perpendicular to the foliation. A) Fine-grained epidote amphibolite with the assemblage hornblende – epidote – plagioclase – magnetite – titanite. It has a peculiar folded layer made mostly of epidote and titanite, which may be some sort of deformed vein or the remains of a large Fe-Ti oxide phenocryst. The foliation distant from the folded layer (red lines) has an orientation somewhat different than the fold axial surfaces (blue line). This suggests that the vein was stronger than the enclosing rock, and either modified the strain field around the layer during deformation, or preserved an earlier foliation. Kvithylla, western Trondheimsfjord, Sør Trøndelag, Norway. B) Detached fold hinge made of polycrystalline plagioclase (partly converted to cloudy retrograde sericite) in an upper amphibolite facies mylonite that had a diorite protolith. The fold axial surface is approximately parallel to the rock foliation (red line). The folded plagioclase layer may have been derived from an extended, recrystallized plagioclase porphyroclast, which are abundant in this rock. Stafford, Connecticut, USA. C) Folded quartz layer surrounding a garnet porphyroclast in a kyanite- and rutile-bearing bearing mylonite. Fjørtoft, Nordøyane, Møre og Romsdal, Norway. D) Asymmetrically folded, alternating quartz-feldspar-rich and biotite-rich layers in mylonitic gneiss. Fold asymmetry indicates that the shear sense is dextral (top-right). Helland, Brattvåg Harbor, southern Moldefjord, Møre og Romsdal, Norway. Abbreviations: B, biotite; G, garnet; H, hornblende; P, plagioclase; Q, quartz; QF, quartz and feldspar; T, titanite.

Figure 16.3 Open fold in layered slate. There is a distinct axial planar cleavage parallel to the red line that cuts this synformal structure. The brown-weathering layer, about midway in the sequence (yellow arrow), is made of carbonate-bearing slate, as are some thinner layers. Dissolution of carbonate causes this layer to weather faster that those above and below. The presence or absence of carbonate may indicate changes in the depositional environment of the original sedimentary basin, for example deposition rate or carbonate compensation depth. A geologist working in metamorphic rocks should think about the protoliths, and what they record in terms of pre-metamorphic geology. Castleton, Vermont, USA.

Figure 16.4 Folded granitic gneiss that has a prominent early foliation (compositional layering and alignment of biotite) that developed in an earlier episode of deformation. The rocks were folded into the form seen here, with little development of axial planar cleavage. Brødreskift, outer Trondheimsfjord, Sør Trøndelag, Norway.

Figure 16.5 More or less symmetrical, similar folds in migmatitic biotite schist. Light-colored leucosomes highlight the folds and parallel an early foliation. A somewhat boudinaged pegmatite dike has cut approximately along one axial surface from bottom center to the upper right (red arrow). The symmetry of these folds, in this intensely deformed region, implies that this location may be near the axial surface of a much larger-scale fold. Higganum, Connecticut, USA.

Figure 16.6 The hinge region of a large recumbent fold on the side of a mountain. The rock face includes a variety of gneisses and schists. The axial surface to this fold is nearly horizontal, and the fold axis extends deeper into the rock face from left to right. Rekdal, Moldefjord, Møre og Romsdal, Norway.

Figure 16.7 Folded granitic gneiss, with layers defined mostly by biotite abundance (biotite-rich layers are darker). A relatively thick, homogeneous layer is highlighted with a red line to help the eye follow the folds. Note how some layers maintain thickness through the folds, and others change thickness. Galgeneset, near Rissa, outer Trondheimsfjord, Sør Trøndelag, Norway.

Figure 16.8 Hinge region of a recumbent isoclinal fold in layered muscovite schist and quartzite. The axial surface of this fold dips gently downward to the right and into the cliff. A graded quartzite layer has a sharp, white base (red arrows) and a gradational top, showing that this structure is an overturned syncline (see Figs. 3.7, 23.9, 23.10). Jaffrey, New Hampshire, USA.

Figure 16.9 Recumbent isoclinal fold in layered quartzite. Red arrows point to two of the several places where blocks of rock in the fold hinges have fallen out, exposing the curved hinge surfaces. The blocks fell out because the dominant rock foliation and cleavage is defined by compositional layering and strongly aligned muscovite. Where muscovite is abundant, the strong cleavage allows blocks to detach easily. Indeed, this is a flagstone quarry, mined specifically because, away from fold hinges, the rock breaks easily along the cleavage into thin, flat slabs. Engan quarry, Sør Trøndelag, Norway.

Figure 16.10 Tight folds in interlayered quartzite and amphibolite. The prominent antiformal hinges (red arrows point out two) have nearly horizontal hinge lines that, on this gently sloping surface, make the folds look more nearly isoclinal than they actually are. A set of synformal hinges are visible to the left of the hammer, between the yellow arrows. Skår, western Moldefjord, Møre og Romsdal, Norway.

Figure 16.11 The dark rocks to the upper left and lower right are basalt dikes, metamorphosed to amphibolite facies conditions. The light-colored rock in between is granitic gneiss. The dike contacts clearly cut the gneissic layering. Although the original gneiss was certainly folded prior to dike intrusion, gneisses distant from this zone of extreme deformation are relatively coarse and not obviously tightly folded. The gneiss seen here is relatively fine-grained and has nearly isoclinal folds that have axial surfaces approximately parallel to the dike contacts. The folding and grain-size reduction resulted from intense shear, approximately parallel to the dike contacts. During deformation, the gneiss was apparently much more ductile than the adjacent amphibolites. Oksvollneset, Midøy, Møre og Romsdal, Norway.

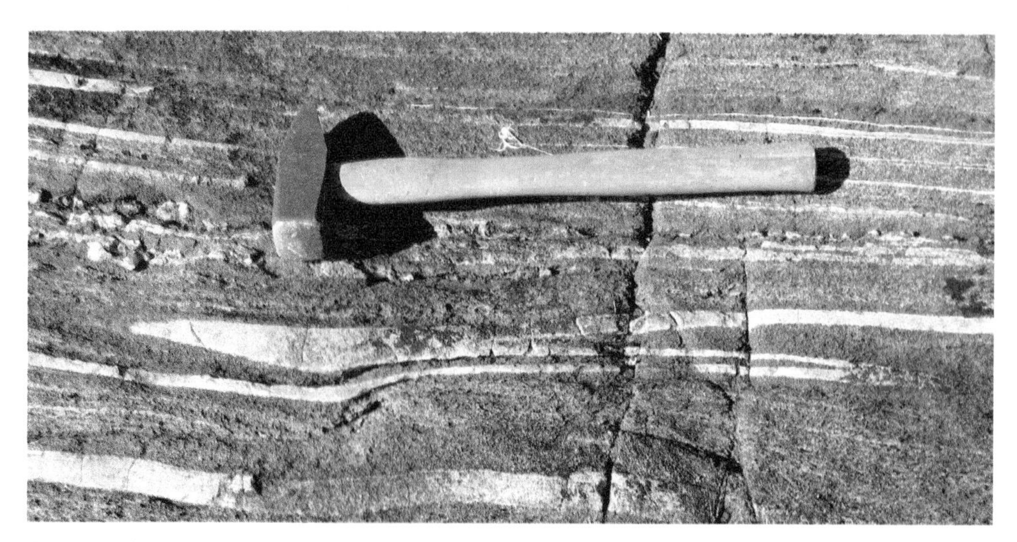

Figure 16.12 Here, white layers of felsic rock occur within darker gray gneiss. The white layer below the hammer has been deformed into a nearly isoclinal, asymmetric fold. Fold asymmetry indicates sinistral shear across this layer. Skardsøya, Møre og Romsdal, Norway.

Figure 16.13 Asymmetrically folded amphibolite dike in layered gneiss. Fold asymmetry indicates dextral shear across this layer. Although the crosscutting relationships are not obvious, it does crosscut contacts between compositionally different gneissic rocks. Quabbin Reservoir, Massachusetts, USA.

Figure 16.14 This vertical surface shows a thin, white layer, within a dark amphibolite layer, between two thick layers of light-gray felsic gneiss. In this case, the relatively ductile amphibolite accommodated much of the strain between the gneiss layers during deformation, resulting in top-left asymmetric folds in the white layer. Ramsvika, Sør Trøndelag, Norway.

Figure 16.15 A small dome produced by the interference of two fold sets. This fine-grained muscovite-garnet schist has a foliation and cleavage approximately parallel to the rock surface, produced during an early episode of deformation. A second episode of deformation produced folds having axial surfaces nearly perpendicular to the early cleavage. The faint intersection lineation of the second cleavage on the early cleavage surface is parallel to the yellow line. A third episode of deformation produced folds with axial surfaces also at nearly right angles to the early cleavage. The faint intersection lineation of the third cleavage on the early foliation surface is parallel to the red line. Constructive interference of two small antiforms of the two late fold generations produced this small dome in the early cleavage surface. Whately, Massachusetts, USA.

Figure 16.16 Relatively simple refolded fold. The dark-gray rock is granitic gneiss that hosts numerous pink pegmatite dikes and boudins. One long pegmatite dike, which presumably was originally a flat feature, was isoclinally folded twice. The first fold episode produced the fold axial surface shown with a dashed red line, with the fold hinge indicated by a red arrow. The original red axial surface was also probably relatively flat, with the pegmatite layer doubled over. The folded pegmatite and the red axial surface were folded in a second episode to produce a new, nearly flat fold axial surface (yellow line), two new fold hinges (yellow arrows), and caused the pegmatite layer to be repeated four times along a vertical line. Fowler, Adirondacks, New York, USA.

Figure 16.17 Folded amphibolites and migmatitic felsic gneiss. The later and most obvious fold that forms the arch has an axial surface parallel to the pen-like magnet. Notice the tight ptygmatic-type folding of a coarse-grained, early pegmatite dike (yellow arrow), with fold axial surfaces parallel to the large, late fold. A late pegmatite dike (red arrows) cuts the left-hand fold limb, to the left of the magnet, and continues upwards close to the fold axial surface. Less obvious are earlier isoclinal fold hinges, the clearest of which are pointed out with blue arrows. Quabbin Reservoir, Massachusetts, USA.

Figure 16.18 Complexly folded granitic gneiss, pegmatite, and biotite schist. The largest, most obvious folds have axial surfaces indicated by the red dashed line. Within the late fold limbs are earlier isoclinal fold hinges. One synformal fold hinge is indicated by a yellow arrow, and two antiformal fold hinges are indicated with blue arrows. The rock layer at the antiformal fold hinges can be traced around a closed region (blue dotted outline), somewhat shaped like a boomerang). Closure of the dotted blue line shows that the two antiformal fold hinges are in fact the same hinge, which curves in three dimensions outside the exposed rock surface (blue dashed line). Fowler, Adirondacks, New York, USA.

Figure 16.19 Ptygmatic folds of thin, white pegmatite dikes that cut the layering in felsic gneiss. Notice that the ptygmatic folding style of the white dikes contrasts with comparatively open, more or less similar folding style of the host gneiss. Hellevik, Lepsøy, Møre og Romsdal, Norway.

Figure 16.20 Ptygmatic folds of pegmatitic leucosomes in a granitic gneiss. Notice the thin, black, biotite-rich selvages on the margins of the pegmatite segregations (some indicated by red arrows). Lensvik, Sør Trøndelag, Norway.

Figure 16.21 Marble with a calc-silicate layer forming a complex closed loop, a sheath fold, most easily seen by tracing the rusty-brown layer immediately surrounding the felt-tipped marker pen. The calc-silicate layer is weathered out in raised relief because of the greater solubility of the surrounding calcite. The resulting 3-dimensional view indicates that the layer surfaces are approximately parallel. Warrensburg, Adirondacks, New York, USA.

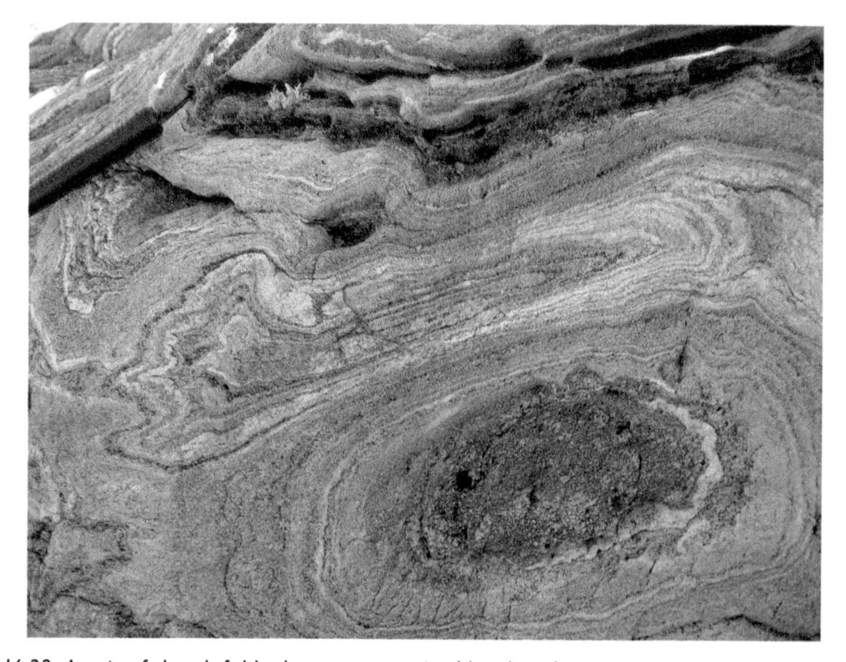

Figure 16.22 A pair of sheath folds that are recognized by closed or concentric compositional layers, forming a coupled pair, one antiformal and the other synformal. In this view it is not clear which is which. The darker rock is amphibolite, which is surrounded by migmatitic felsic gneiss. Ullaholmen, Nordøyane, Møre og Romsdal, Norway.

Chapter 17

Porphyroblasts, porphyroclasts, and augen

Some igneous rocks contain phenocrysts, which are large crystals that grew during an initial period of slow crystallization, set in a comparatively much finer-grained matrix that crystallized in a later phase of more rapid crystallization. Metamorphic rocks have somewhat similar things, large crystals set in a finer-grained matrix, called porphyroblasts. These, however, grew in solid rock rather than a silicate liquid. Small numbers of large crystals can grow during metamorphism if a new mineral nucleates (starts to grow) in only a few places, under conditions where the chemical components needed for the new crystals can migrate quickly. This may be accomplished, for example, by diffusion through or flow of grain boundary fluid. Loss of those chemical components from the fluid to the growing porphyroblasts limits the activity (effective concentration) of those components, which limits the degree of supersaturation, which in turn inhibits nucleation of new grains. Ultimately, the small number of mineral nuclei grow into a small number of large crystals.

Porphyroblasts can be euhedral, with well-formed crystal shapes, or they can be subhedral or anhedral, having few or no recognizable crystal facets. Because they grew in solid rock, they can overgrow adjacent minerals, engulfing them as inclusions, or they can push the minerals aside. In some cases, both can occur along different faces of the same crystal. Inclusions can preserve samples of mineral assemblages from early parts of the metamorphic history, that may have been lost outside of the porphyroblasts. For example, chloritoid preserved only in garnet crystals in a staurolite-kyanite schist (Fig. 1.5), or rutile preserved only in garnet in a retrograded eclogite. Mineral inclusions from earlier metamorphic assemblages can help reconstruct the pressure-temperature paths that rocks took during metamorphism. Porphyroblasts may overgrow the rock foliation, preserving it and showing how it changed during structural development. In this way, porphyroblasts can preserve structural information that may not be available elsewhere (e.g., Fig. 10.1A, 15.2A).

Porphyroclasts differ from porphyroblasts in that they are notably large crystals, or in some cases coarse, polycrystalline masses, that have been deformed by fracture, recrystallization, or some other process. Porphyroclasts do not form exclusively from metamorphic porphyroblasts, but can also be remnants of igneous phenocrysts from the original protolith, or coarse crystals from disrupted pegmatites or veins. The important points are that the large crystals are set in a finer-grained matrix, and the crystals no longer have their original shapes.

Porphyroclast deformation styles and mechanisms can give information on the metamorphic conditions, strain rate, and local shear sense when the rock was deformed.

Figure 17.1 shows some schematic examples of porphyroclasts that developed from two example large starting crystals: a metamorphic porphyroblast (A) and a phenocryst (B). A and B can potentially deform into any of the resulting porphyroclasts illustrated in C-F, which differ in deformational style: fracture during shear (C), recrystallization during flattening (D), recrystallization during shear (E), and recrystallization during shear with porphyroclast rotation (F).

The thin section images show some examples of porphyroblasts (Fig. 17.2A-C) and porphyroclasts (Fig. 17.2D-F). The field photos are similarly organized, starting with porphyroblasts and some associated textures in Figures 17.3–17.9. Figures 17.10–17.12 show a sequence from metamorphosed but essentially undeformed porphyritic granite to porphyroclastic augen gneiss. Figures 17.13–17.15 show boudins and porphyroclasts derived from disruption of granitic pegmatites, that can ultimately result in complete separation of large crystals from one another. Figures 17.16 and 17.17 show porphyroclasts of uncertain origin, but probably from disruption of larger bodies, analogous to the granitic pegmatites.

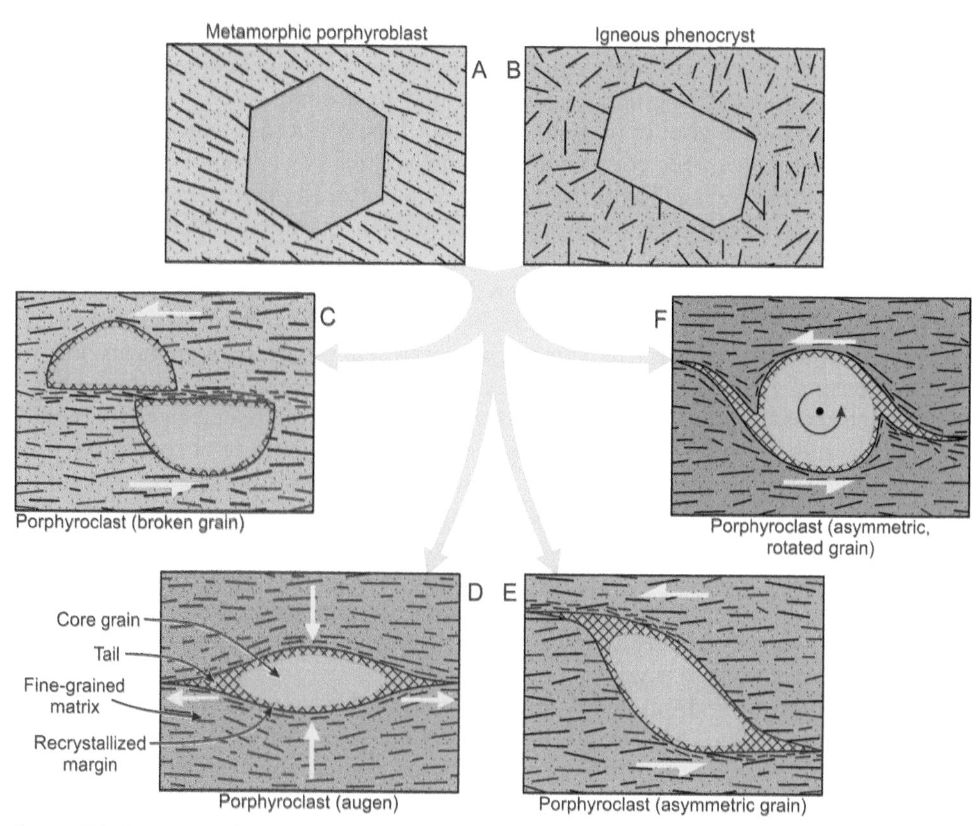

Figure 17.1 Examples of a metamorphic porphyroblast (A), and an igneous phenocryst (B) prior to deformation. In C-F the precursor crystals have been transformed to porphyroclasts by shape change during metamorphic deformation. Porphyroclasts can develop through a variety of deformational styles (e.g., Passchier and Simpson, 1986). C) Broken grains. D) Recrystallization to produce symmetric, eye-shaped grains (augen, symmetric shape implies the importance of flattening during deformation). E) Recrystallization to produce asymmetrically deformed grains (sigma porphyroclast, in this case indicating a sinistral shear sense). F) Recrystallization to produce asymmetrically deformed grains, with subsequent grain rotation (delta porphyroclast, in this case also indicating a sinistral shear sense).

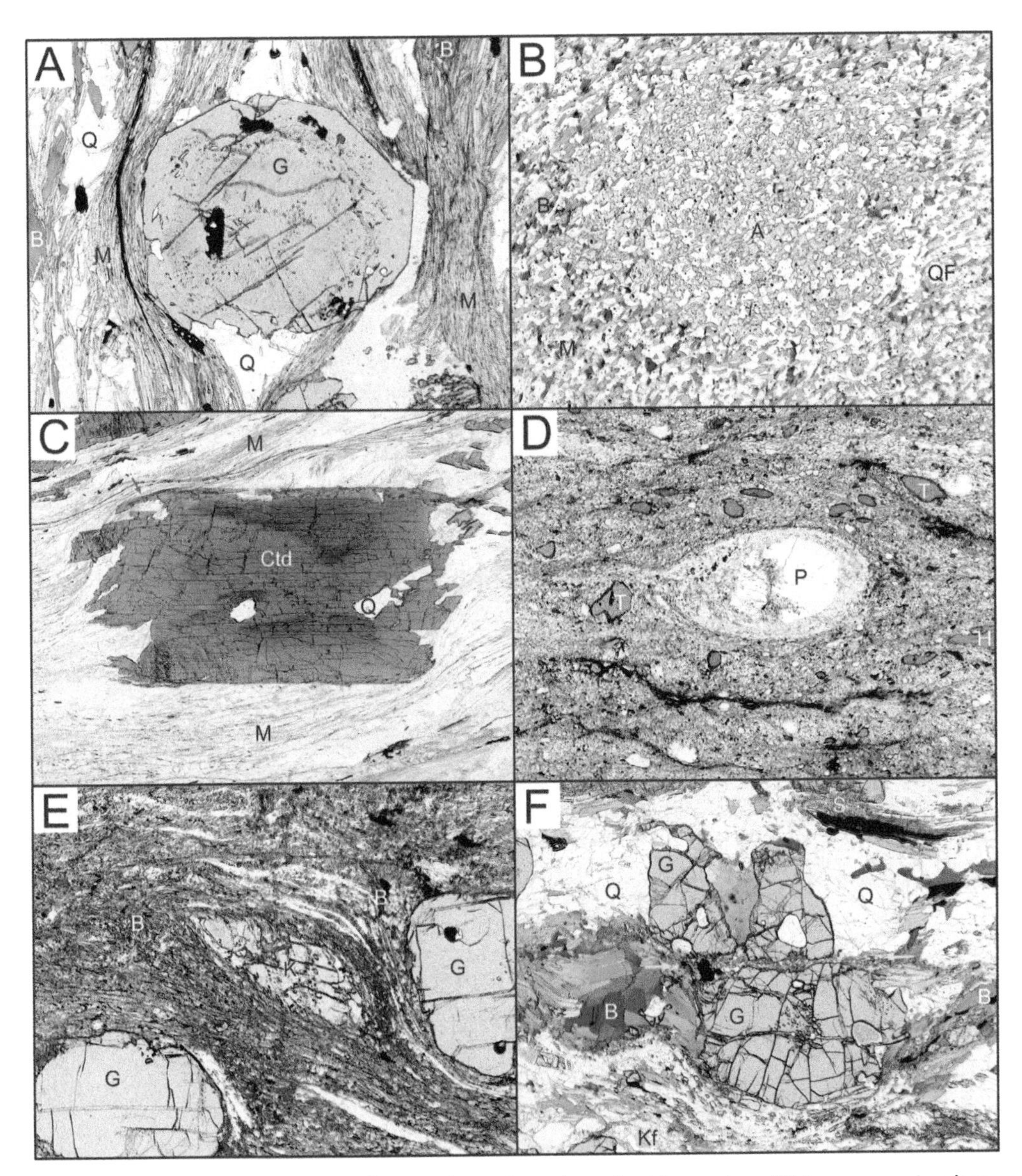

Figure 17.2 Examples of porphyroblasts and porphyroclasts in thin section. All images are in plane-polarized light, and have field widths of 4 mm. A) Garnet porphyroblast in a sillimanite-staurolite grade pelitic schist. Orange, Massachusetts, USA. B) Contact metamorphosed slate, now granofels, showing an anhedral, inclusion-rich (poikilitic) porphyroblast of andalusite (center) in a matrix of quartz, feldspar, and biotite (like Fig. 13.6). Greenville Junction, Maine, USA. C) Chloritoid porphyroblast in chlorite grade phyllite. Initial stages of deformation of one corner of the chloritoid crystal is indicated by a red arrow. Notice the sector color zoning. Tyson, Vermont, USA. D) Deformed diorite with porphyroclasts of three original igneous minerals: plagioclase (center), titanite, and hornblende. The clear center of the plagioclase is a core single crystal, surrounded by an outer, cloudy layer of small, recrystallized plagioclase grains. At first glance this looks like a symmetrical augen, but a slight asymmetry suggests a sinistral shear sense. Stafford, Connecticut, USA. E) Mylonitic pelitic schist with almost unmodified garnet porphyroblasts and an asymmetric kyanite porphyroclast in the center. Asymmetry indicates a sinistral shear sense. The outer portion of the kyanite is partially replaced by needle-shaped sillimanite. Fjørtoft, Nordøyane, Møre og Romsdal, Norway. F) A garnet – cordierite – sillimanite – biotite pelitic schist containing a garnet

Figure 17.2 (Continued) that was fractured (between yellow arrows) into two pieces to form two porphyroclasts. The shear zone is in the same orientation as the rock foliation and cleavage. The offset direction along the fracture indicates sinistral shear sense. Sturbridge, Massachusetts, USA. Abbreviations: A, andalusite; B, biotite; Ctd, chloritoid; G, garnet; H, hornblende; K, kyanite; Kf, K-feldspar; M, muscovite; P, plagioclase, Q, quartz; QF, quartz and feldspar; T, titanite.

Figure 17.3 Garnet porphyroblasts in an amphibolite. The rock has a weak foliation oriented vertically, and the garnets are mostly euhedral to subhedral. Some of the garnets have concentric color zoning, caused by differing amounts of mineral inclusions. Midsund, Otrøy, Møre og Romsdal, Norway.

Figure 17.4 Eight centimeter long kyanite porphyroblast in a partially retrograded kyanite eclogite. Most kyanite crystals in this rock are only 5 mm long or less. Why this particular grain is so large is not clear. Gossa Island, Møre og Romsdal, Norway.

Figure 17.5 Porphyroblasts of andalusite (now pseudomorphed by sillimanite) in schist (upper half of the photo) and impure quartzite (lower half). Many of these porphyroblasts cross the mica foliation, oriented parallel to the red line, suggesting that they grew relatively late in the deformation history of this rock. Jaffrey, New Hampshire, USA.

Figure 17.6 This is a single crystal of garnet that is part of a vermicular intergrowth with K-feldspar. This remarkable porphyroblast, and its neighbors, are associated with disrupted pegmatites in a sillimanite – K-feldspar grade sillimanite – biotite – garnet gneiss. One idea for the origin of this texture is that it represents the prograde breakdown products of original large mica crystals in the pegmatite, for example: biotite + muscovite + 3quartz = garnet + 2K-feldspar + $2H_2O$, or biotite + sillimanite + 2quartz = garnet + K-feldspar + H_2O. Comstock, eastern Adirondacks, New York, USA.

Figure 17.7 Fine-grained muscovite – biotite schist with large, euhedral porphyroblasts of brown staurolite. The staurolite crystals are approximately aligned with the rock foliation and cleavage (parallel to the yellow line). The cleavage wraps around some of the larger porphyroblasts, which is particularly clear near the right end of the largest crystal. Some of the staurolite crystals have thin muscovite-rich rims (red arrows) that result from a retrograde hydration reaction. Montcalm, New Hampshire, USA.

Figure 17.8 This is the same fine-grained muscovite – biotite schist as Figure 17.7, but from a greater distance and showing broken rock cleavage faces. Few staurolite porphyroblasts are visible, but instead there are myriad bumps and pits where the rock cleavage has broken around the large staurolite crystals. The rectangular shapes of some bumps and pits, where the rock cleaved close to the staurolite crystal surfaces, can be seen (red arrows). Trying to break rocks like this, to expose the porphyroblasts, can just make a powdery mess. It is sometimes necessary to cut the rock with a diamond saw to expose them well. Montcalm, New Hampshire, USA.

Figure 17.9 Not all porphyroblasts are large, they just have to be much larger than the matrix crystals. This is an example of small porphyroblasts of hornblende in a rusty-weathering felsic gneiss. Note that most of the hornblende crystals are parallel to the foliation, but some cross it. The crosscutting hornblende crystals may be early, not having been fully rotated into parallelism with the foliation, or they may have grown late, after deformation had largely ceased. Rissa, outer Trondheimsfjord, Sør Trøndelag, Norway.

Figure 17.10 This is an essentially undeformed block of phenocryst-rich rapakivi granite, metamorphosed at amphibolite facies conditions. The rapakivi texture is evident from the yellowish-white albite rims (yellow arrow) surrounding some of the pinkish-gray K-feldspar crystal cores (red arrow). This block is surrounded by more severely deformed gneiss that is poorer in original phenocrysts. The reason this block resisted deformation is probably the high proportion of strong, deformation-resistant, coarse feldspars. Figures 17.11 and 17.12 are of gneisses derived from this same unit at about the same metamorphic grade, though from rocks that had smaller proportions of phenocrysts than this one. Skår, Moldefjord, Møre og Romsdal, Norway.

Figure 17.11 Mylonitic gneiss, derived from the same unit as Figure 17.10, but having had fewer original phenocrysts. Here the phenocrysts have been deformed, reduced in size, and in some cases broken, all forming metamorphic porphyroclasts. Most of the porphyroclasts are aligned with the foliation, which is approximately parallel to the hammer head. Movement sense is sinistral, as indicated by most porphyroclast asymmetries. Lauvøy, Moldefjord, Møre og Romsdal, Norway.

Figure 17.12 This rock has also been derived from the same unit as Figure 17.10. Here the porphyroclasts have been deformed into mostly eye-shaped augen. Movement sense here is dextral, as indicated by asymmetries on some of the porphyroclasts. 10 km South of Engan, Sør Trøndelag, Norway.

Figure 17.13 Gray granodioritic gneiss metamorphosed to amphibolite facies conditions, containing a severely deformed, largely disrupted pegmatite. There are polycrystalline boudins of white albite and pink microcline, the largest of which are indicated by blue arrows. There are also single crystal porphyroclasts that have been mostly separated from their original pegmatitic neighbors, some of which are indicated by red arrows. Under the metamorphic conditions during deformation, the feldspars remained relatively intact compared to quartz and micas, which were recrystallized into fine-grained obscurity. With sufficient deformation, crystals from disrupted pegmatites can become completely isolated, with no obvious connection to their pegmatitic origins. Chester, Connecticut, USA.

Figure 17.14 Feldspar porphyroclasts in mylonite, showing several recrystallized feldspar tails and a confusing set of shear sense indicators. The smaller porphyroclasts (two indicated with red arrows) generally indicate a dextral shear sense. The larger porphyroclasts in a row across the image center indicate the same shear sense, if they are interpreted as having rotated (delta porphyroclasts, Fig. 17.1F). Midsund, Midøy, Møre og Romsdal, Norway.

Figure 17.15 This is a single crystal of white microcline in garnet – sillimanite – biotite gneiss. Looking at this porphyroclast in isolation, its origin might be unclear. In the same outcrop, however, there is every gradation between deformed but otherwise obvious pegmatites, to pegmatite boudins, to strings of crystals, to isolated crystals like this. It is clear from the progression that these isolated crystals are not deformed porphyroblasts or phenocrysts, but pieces of disrupted pegmatite. Comstock, eastern Adirondacks, New York, USA.

Figure 17.16 Pyrite porphyroclast in amphibolite (quartz vein to the lower left, asymmetric folds to the upper right). One of the tails of this porphyroclast can be seen extending from the pyrite grain to the lower-right corner of the image, where the pyrite has weathered out from the outcrop surface. This may be a deformed pyrite porphyroblast but, considering the sulfide mineralization in nearby rocks, it may instead be part of a disrupted sulfide vein. If so, this grain might be analogous to the disrupted pegmatites in Figures 17.13–17.15. Surnadal, Møre og Romsdal, Norway

Figure 17.17 Pyrope porphyroclast in garnet peridotite. These large crystals are widely separated in this mantle-derived ultramafic body. The elongate shape of this garnet crystal, compared to much smaller, round garnet crystals in the matrix, suggests that it is a porphyroclast. By analogy with Figures 17.13–17.16, these magnificent crystals may have been derived from disruption of pegmatite-like bodies in the mantle, possibly precipitated from fluid or basaltic liquid migrating along fractures. Ugelvik, Ottrøy, Møre og Romsdal, Norway.

Chapter 18

Boudins

A boudin is an elongate fragment of a layer (or dike or vein) that has become partially or completely detached during extension. The word 'boudin' comes from French, for sausage ( ), referring to the typically elongated, sausage-like shape of these fragments (Fig. 18.1). Boudins form by the extensional disruption of stronger layers within more ductile rocks. The line along which layer fragmentation occurs is commonly periodic, giving boudins a regular spacing. The competent layers (Fig. 18.1C) tend to break into segments in a brittle (Fig. 18.1D, H), semi-brittle (Fig. 18.1A, E), or ductile way (Fig. 18.1F, G). The more ductile enclosing rocks can deform to fill the potential void spaces between the boudins. In some cases, fluids or melts migrate into the potential void spaces, eventually filling them with solid material such as quartz or pegmatite (Fig. 18.1E).

The combination of shearing and extensional strain can asymmetrically deform or rotate the boudin segments with respect to one another. This gives the boudins an asymmetry that can be used as shear sense indicators for both ductile and brittle boudins (Figs. 18.1G and H, respectively). Boudins vary widely in appearance, depending on the amount of deformation, deformation direction relative to the layer orientation, ductility and ductility contrast, shear sense, and the orientation of the outcrop surface with respect to the boudins (e.g., Goscombe et al., 2004).

The field images are here sorted in order from brittle (Figs. 18.2–18.4), to transitional between brittle and ductile (Figs. 18.5–18.7), to ductile (Figs. 18.8–18.10). Figure 18.11 shows a boudin that is almost round, having been distantly separated from neighboring boudins, and having lost its boudin necks. Figure 18.12 shows how mullions can develop on a cleavage surface, as the result of boudinaged layers below that surface. Figures 18.13 and 18.14 show asymmetric boudins that developed during shear. Finally, Figure 18.15 shows a subtle boudin as a reminder that nature tends to cover many geologic features, making them more difficult to spot. Think of it as a practice example, to help readers train their eyes to extract patterns from beneath the obscuring cover.

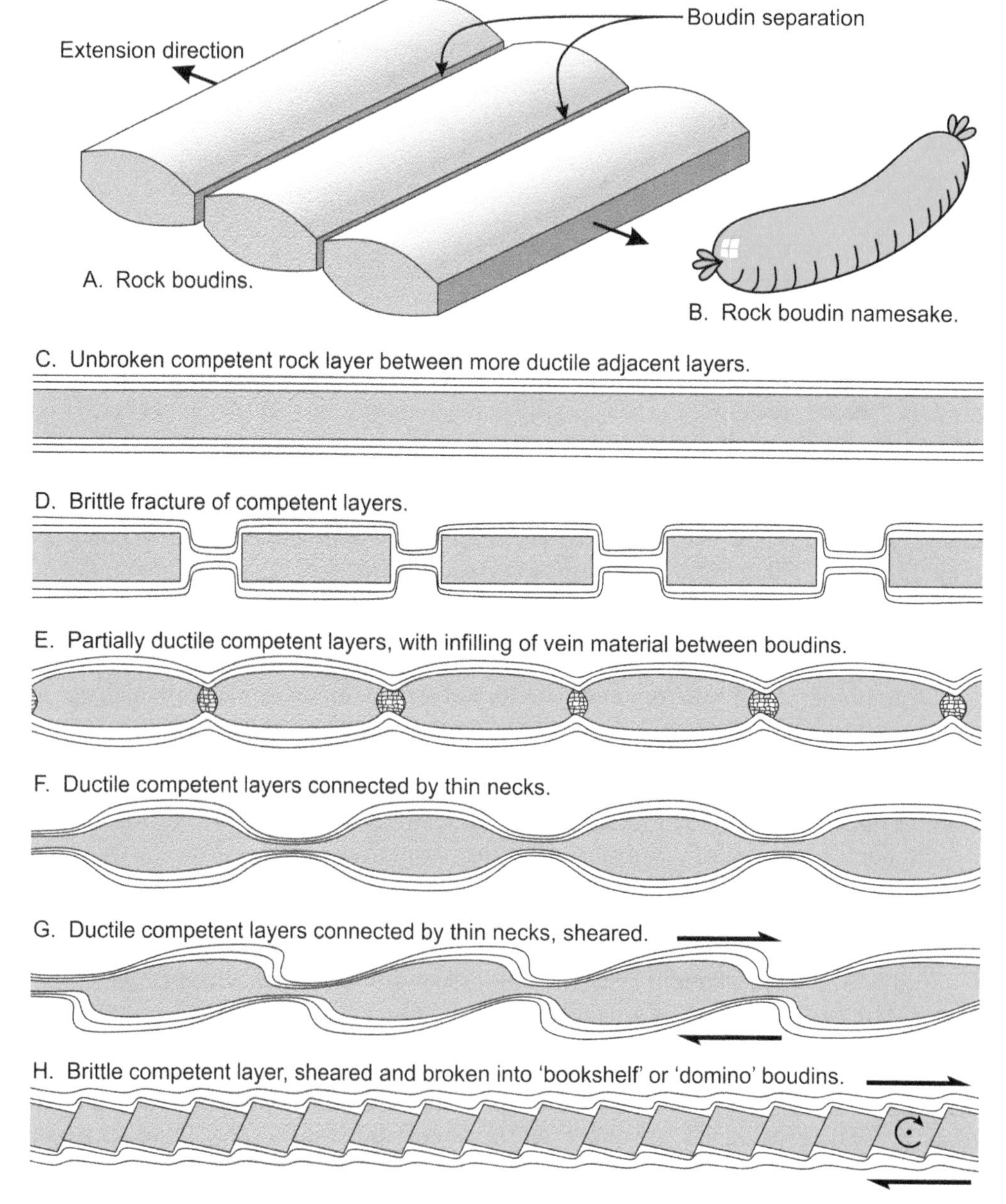

Figure 18.1 Schematic examples of boudins. A) The elongated shape of boudins vaguely resembles sausages, (B). They form from a competent layer that has been extended, with boudin separation or neck lines perpendicular to the extension direction. C) Unbroken competent layer. D) Brittle boudins. E) Semi-brittle boudins with infilling of the boudin neck regions with vein material, precipitated from metamorphic fluid. F) Ductile boudins, showing thinning of the boudin necks, but not breakage. G) Ductile boudins that extended during dextral (top-right) shear. H) Brittle 'bookshelf' or 'domino' boudins that developed during extension and dextral shear, with rotation of the individual boudinaged blocks. Note that D–H are sections perpendicular to the boudin length. A similar cross section of B would be round, which is not typical of rock boudins (but see Fig. 18.11).

Figure 18.2 Boudinaged basaltic dike in highly deformed marble that was metamorphosed to amphibolite facies conditions. The dike broke in a brittle fashion into roughly rectangular blocks. Ductile flow of the folded marble has separated and displaced the dike segments. With the metasomatic addition of calcium from the marble, the dike boudins are now clinopyroxene-bearing amphibolite. Northwest Adirondacks, New York, USA.

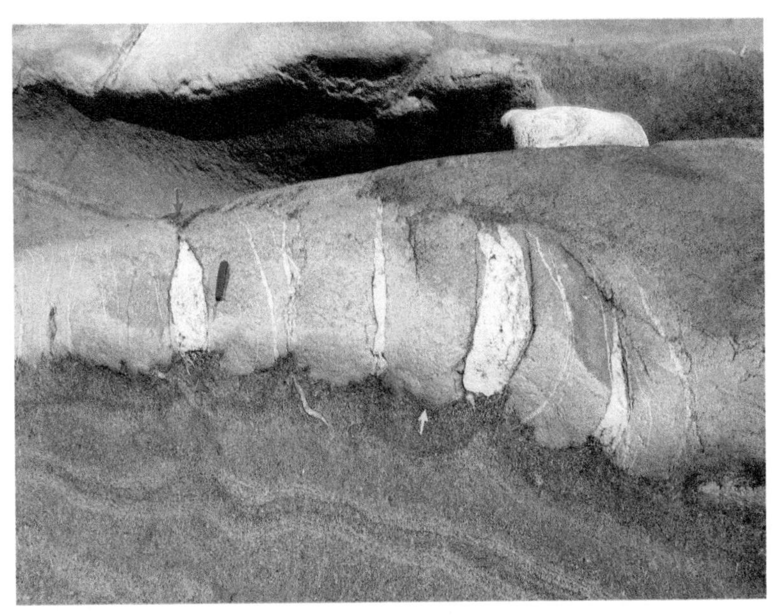

Figure 18.3 This light-gray quartzite layer (between the blue and yellow arrows) was boudinaged when it was slightly ductile, causing the surrounding, more ductile, rusty-weathering muscovite – biotite schist to fold slightly into the intra-boudin potential void spaces (red arrows point out the in-folding in one case). Those spaces were filled by fluid, which precipitated white vein quartz. The quartzite layer originally formed as a turbidite deposit. The yellow arrow points to the sharp contact with schist at the quartzite layer bottom, and the blue arrow points to the more gradational layer top. The layer is therefore right side-up. Huntington, Massachusetts, USA.

Figure 18.4 Layered dolomitic marble (white and tan) spectacularly folded into the somewhat brittle boudin necks of a partially boudinaged basaltic dike (gray). Thompson Belt, Canada.

Figure 18.5 Looking down on a boudinaged amphibolite layer in rusty-weathering schist. The amphibolite has undergone partially ductile deformation, with infolding of schist into the boudin necks, followed by brittle separation. Vein quartz (red arrows) fills the boudin necks where they have separated. Coastal Maine, USA.

Figure 18.6 Boudinaged amphibolite layer between thin, white layers of quartzite and non-boudinaged amphibolite layers. The boudins deformed partly by ductile flow, as indicated by periodic thinning of the boudinaged amphibolite layer, and partly by brittle failure as indicated by the broken boudin ends. Metamorphic fluids migrated into the potential voids where the boudins separated, to precipitate white vein quartz. Hasselvika, Trondheimsfjord, Sør Trøndelag, Norway.

Figure 18.7 Amphibolite boudins in layered and folded, somewhat rusty-weathering garnet-biotite schist. Notice quartz filling the boudin neck region (center), and fractures perpendicular to the apparent extension direction, in the amphibolite, to the left and right of the quartz filling (red arrows point out two). The fractures probably post-date boudin formation. Revsneshagen, outer Trondheimsfjord, Sør Trøndelag, Norway.

Figure 18.8 Boudinaged white pegmatite and black amphibolite, in gray biotite schist. The pegmatite bodies have been partially dismembered into boudins, and no longer form continuous dikes or sills. The large, egg-shaped pegmatite boudin in the image center is about 1.5 m thick. Below and slightly to the right of that boudin is a black amphibolite boudin (B), with a long tail extending to the left to a smaller amphibolite boudin (b). Orkanger, Trondheimsfjord, Sør Trøndelag, Norway.

Figure 18.9 The dark-gray pyroxene – garnet granulite boudin in the lower left is surrounded by pink and light-gray granitic gneiss. During boudin formation, the pyroxene granulite boudin edge stretched out into a thin neck, (arrow) only a few percent the thickness of the boudin body. This boudin neck may have connected to another boudin to the right, that unfortunately has eroded away. Comstock, eastern Adirondacks, New York, USA.

Figure 18.10 Eclogite boudin about 100 m x 150 m (outlined in white) exposed on a fjord wall, surrounded by biotite schist and amphibolite. Boudins that are widely separated or isolated from one another indicate a lot of rock strain. Midsund, Midøy, Møre og Romsdal, Norway.

Figure 18.11 Nearly round cross section of an eclogite boudin, surrounded by garnet-biotite schist. Once part of an extensive layer, this boudin become detached and isolated from the rest. A change in strain orientations after boudin separation has severely attenuated or sheared off the boudin necks, giving this boudin an almost boulder-like appearance (see Fig. 7.1E and F). Notice how the host rock foliation wraps around the boudin. Gossa Island, Møre og Romsdal, Norway.

Figure 18.12 This complexly folded gneiss contains irregular felsic and mafic layers of variable thickness. Some of the felsic layers have been boudinaged. The grooves and ridges that run along the exposed surface to the lower right are lineations called mullions, that result from the cleavage surface wrapping around the boudinaged layers. Mullions can also be produced by the detachment and isolation of thick fold hinges of vein quartz, pegmatite, or similar materials in more ductile rock. The field width is about 2.5 meters. Rissa, western Trondheimsfjord, Sør Trøndelag, Norway.

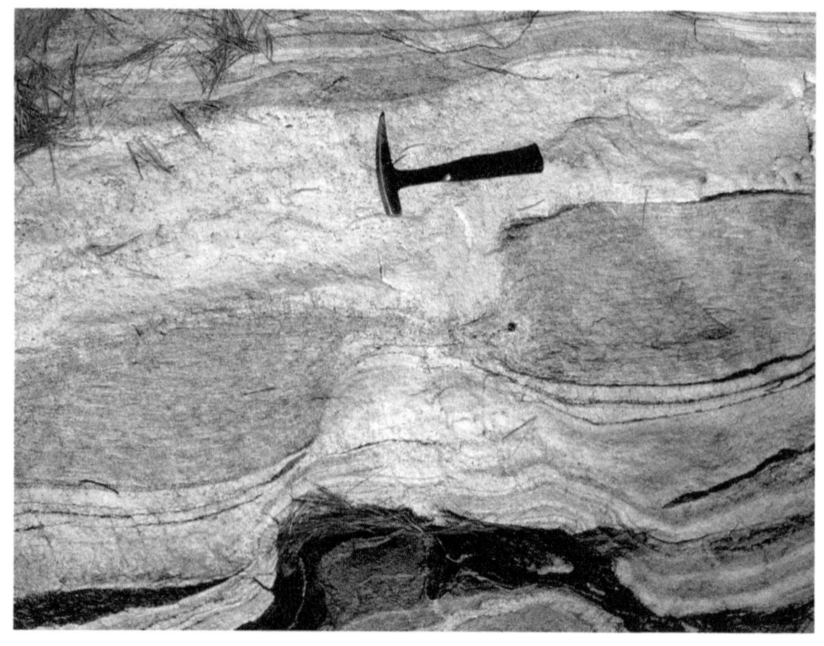

Figure 18.13 A boudinaged layer of gray tonalitic gneiss between layers of more ductile, white gneiss. The two boudins are still connected by a thin neck of gray gneiss. The asymmetry indicates dextral shear. Quabbin Reservoir, central Massachusetts, USA.

Figure 18.14 Domino or bookshelf boudins in a thick amphibolite layer in quartzite. Extension caused brittle fracture of the amphibolite layer, and sinistral shear caused counter-clockwise rotation of the amphibolite blocks. The surfaces between the blocks are effectively microfaults. The thinner amphibolite layers were more ductile and did not form boudins. Helleneset, Moldefjord, More og Romsdal, Norway.

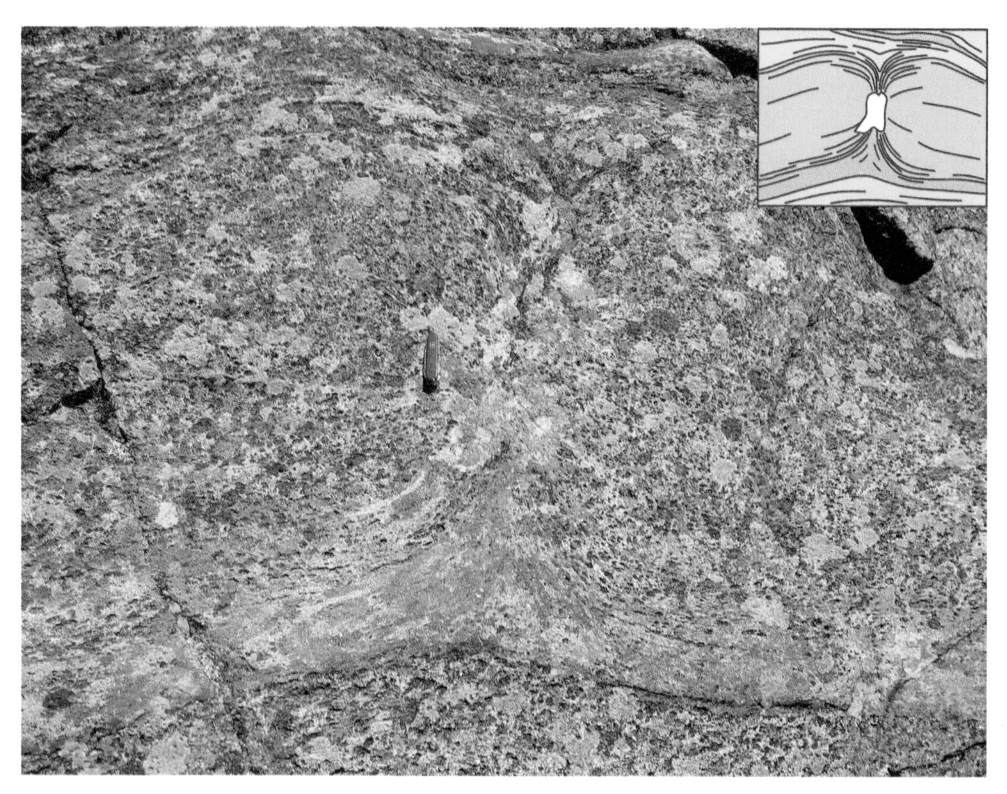

Figure 18.15 Can you see me? Like many things in field geology, boudins can be hard to spot. If it isn't swamps and soil or sand and talus, it's moss and lichen that cover up what you want to see. Honing ones eyes to spot subtle patterns can greatly expand the number of visible geologic features. Here are two boudins of relatively strong schist (blue in the inset), between more ductile layers of quartzite (gray in the inset) and ductile schist (yellow). In the boudin separation region there is a small body of vein quartz (white), precipitated by fluids that migrated into the potential void space as the boudins separated. Lichen, and the rather uniform gray rock color, make the structures hard to spot. With practice looking for such things, you will find that they eventually stand out more clearly. You may also have to practice convincing others that what you see is actually there. Good luck! Jaffrey, New Hampshire, USA.

Chapter 19

Veins and hydrothermal alteration

"In May 1958 altered lavas were dredged by R.R.S. Discovery II from the floor of the deep Atlantic some 450 km west of the coast of Portugal...." "Thin sections of these highly altered rocks are 75 % semi-opaque mess."Matthews (1971)

Veins are mineral deposits left in fractures by flowing fluids. At low pressures, rocks are strong enough to sustain open fractures. Hot water, especially at the higher temperatures achievable at the higher pressures at depth, can dissolve considerable rock material. Open fracture networks permit deeply penetrating surface waters to flow into and out of hot regions by density-driven convection, and other hydrostatic forces. As the fluids migrate, the local conditions of pressure and temperature will change, resulting in the fluid cooling, boiling, mixing with other fluids, and so on. If the fluids become supersaturated with one or more solid phases, minerals will precipitate on the fracture walls. The minerals may grow freely into fluid-filled space, and can fill the space completely. If vein filling is incomplete, the resulting minerals can have fine crystal forms. If a single vein could be followed for a long distance, the precipitated mineral assemblage would probably change along its length, the result of changing conditions and fluid composition as it moved along the fracture.

Fluids in many shallow hydrothermal systems vent to the surface. Places where this happens include the subaerial geothermal fields in Iceland, the Tuscany and Lazio regions of Italy, the Leyte and Negros Islands in the Philippines, the North Island of New Zealand, and many other areas. There are also the remarkable submarine hydrothermal systems and vents that are scattered along the axes of ocean ridges (see Fig. 1.9D).

At deeper levels, rocks are not themselves strong enough to sustain open fractures. Fractures can still form, however, if they become filled with fluid that is close to lithostatic pressure, which occurs if the fractures are not freely open to the surface (Fig. 1.9A). Fluid sources during metamorphism are typically dehydration or decarbonation reactions of hydrous or carbonate-bearing minerals, for example:

$$\underset{\text{chlorite}}{2(Mg_{4.5}Al_{1.5})(Al_{1.5}Si_{2.5})O_{10}(OH)_8} + \underset{\text{quartz}}{4SiO_2} = \underset{\text{garnet}}{3Mg_3Al_2Si_3O12} + \underset{\text{fluid.}}{8H_2O} \qquad (19.1)$$

$$\underset{\text{chlorite}}{(Mg_5Al)(AlSi_3)O_{10}(OH)_8} + \underset{\text{quartz}}{7SiO_2} + \underset{\text{calcite}}{3CaCO_3} = \underset{\text{anorthite}}{CaAl_2Si_2O_8} + \underset{\text{actinolite}}{Ca_2Mg_5Si_8O_{22}(OH)_2} + \underset{\text{fluid.}}{3CO_2 + 3H_2O} \quad (19.2)$$

The first reaction assumes chlorite with three Al per formula unit, similar to that in low-grade pelitic schists, whereas the second reaction assumes two Al per formula unit, more typical of chlorite in calcareous rocks.

Metamorphic fluids are generally thought to initially form an interconnected grain-boundary film. The interconnected fluid film will tend to move down hydraulic pressure gradients, which may be tectonic but will commonly be upwards as a result of the density difference between fluid and the enclosing higher density rock (Fig. 19.1A). The rocks may fracture under hydraulic or tectonic stresses, and the fractures will fill with the available fluid (Fig. 19.1B), at or close to lithostatic pressure. Fluids can migrate long distances through fracture networks, or through rocks having discontinuous fractures and intervening regions of connected grain boundary fluid (Figs. 19.1C, D).

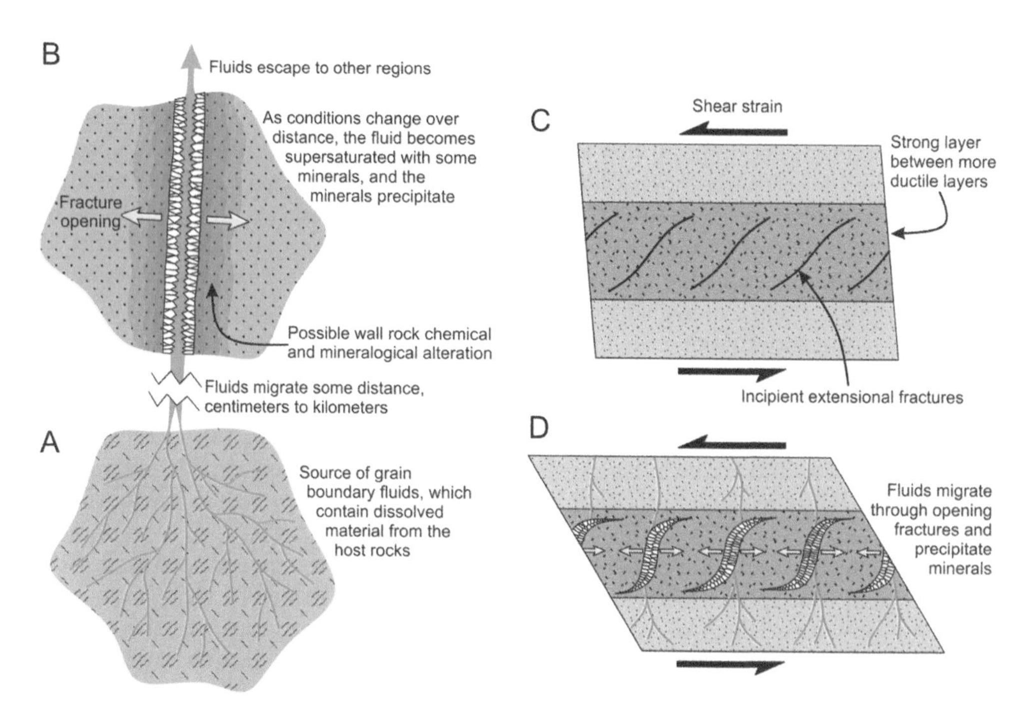

Figure 19.1 Schematics showing two example vein systems. A) The source of fluids during prograde metamorphism is largely dehydration or decarbonation reactions. The initial grain boundary fluids tend to migrate upward because of their low density, eventually coalescing and fracturing their way upwards. As it moves, the fluid encounters lower temperatures or other conditions that cause minerals to precipitate from solution. Other conditions can include lower pressure (especially with boiling), mixing with fluids of different compositions, or reactions with wall rock. Precipitation gradually forms mineral-filled veins (B). C) A competent layer, between two more ductile layers, fractures as it deforms. As the fractures open they form low pressure regions (potential voids) into which grain boundary fluid flows. The fractures provide low-resistance channels, promoting continued fluid flow (D). Unless the fluid is silicate melt, the dissolved solid fraction of the fluid initially filling the fracture is much too low to completely, or even largely, fill the opening space with precipitated minerals. That probably requires many fracture volumes of flowing fluid.

Figure 19.2 shows thin section photomicrographs of veins in metamorphic rocks. The first two rocks (Figs. 19.2A-D) formed at relatively low temperature and have highly imperfect quartz grains that have experienced little or no recrystallization. Figures 19.2E and F show a higher-temperature quartz – tourmaline vein in which the quartz grains are sharply defined, possibly recrystallized following emplacement. It was mentioned in Chapter 1 that metamorphism is traditionally limited to temperatures above 200°C. However, the first few field images given below show shallow features that are the lower-temperature, upper parts of hydrothermal systems (Figs. 19.3, 19.4). These were included to emphasize the connection between the surface manifestations of hydrothermal systems, and veins that are inferred to form deeper down. The next set of field images show hydrothermal alteration and veins that formed at relatively

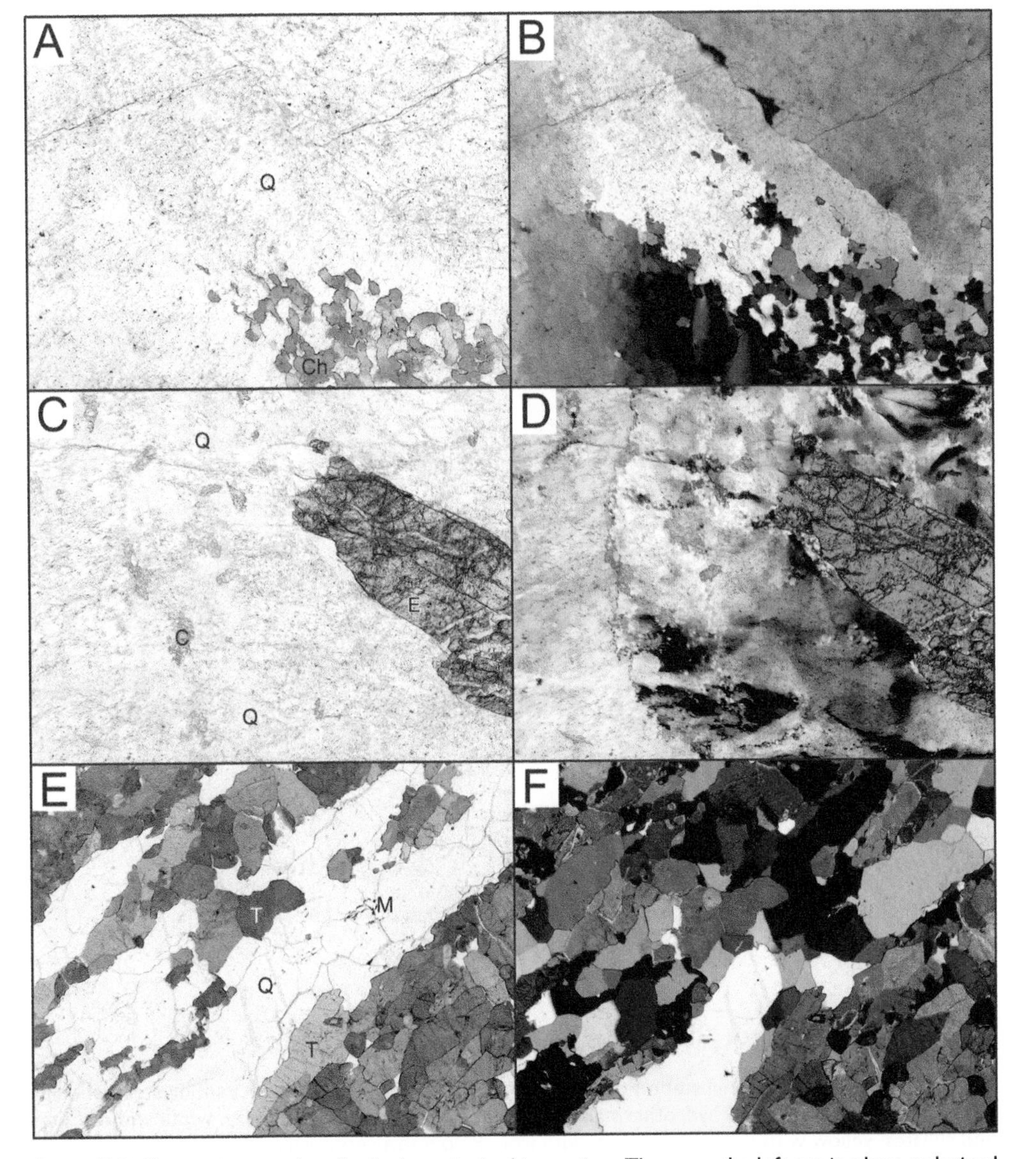

Figure 19.2 Photomicrographs of vein deposits in thin section. Those on the left are in plane-polarized light, those on the right are the same fields in cross-polarized light. All field widths are 4 mm.

Figure 19.2 (Continued) A, B) Quartz vein in chlorite grade phyllite. Greenish, vermicular grains in A (blue in B) are chlorite. The quartz is cloudy because of enormous numbers of fluid inclusions along healed fractures. In B one can see that the quartz is highly strained, with undulatory extinction and polygonal domain textures. Petersburg, New York, USA. C, D) Quartz – epidote – calcite vein that cuts greenschist facies basaltic rocks. Although the adjacent greenschists did not seem to contain much epidote, the quartz vein does. The quartz is cloudy both because of fluid inclusions in myriad healed fractures, and many small calcite crystals. In D one can see that the quartz is highly strained, with undulatory extinction and sutured grain boundaries. Littleton, New Hampshire, USA. E, F) Quartz – tourmaline vein that cuts sillimanite – muscovite grade pelitic schist. A nearby metamorphosed but undeformed dioritic dike, inferred to have the same age as the vein (Thompson, 1988), has an epidote amphibolite facies assemblage, suggesting higher temperatures at the time of vein emplacement than those in A-D. Recrystallization at the higher emplacement temperature may explain the nearly strain-free, equant, clear quartz in E-F. Jaffrey, New Hampshire, USA. The fluids that deposited veins shown in A-D were probably derived from the metamorphic dehydration of local rocks. The vein in E-F appears to have been derived from fluids released from a nearby tourmaline-bearing granite pluton. C, calcite; E, epidote; M, muscovite; Q, quartz; T, tourmaline.

low temperatures, though possibly above 200°C (Figs. 19.5–19.9). Figures 19.10–19.12 show quartz veins that formed in metamorphic rock, probably at or close to lithostatic pressure. Figures 19.13–19.16 show more complex veins that also probably formed at high temperatures and at or close to lithostatic pressure. Lastly, Figure 19.17 shows a boudinaged piece of a high-temperature sulfide-rich mineral vein, as a reminder of the relationship between metamorphic rocks, hydrothermal veins, and many ore deposits.

Figure 19.3 The surface manifestation of a shallow hydrothermal system, with venting steam, boiling mud pots (foreground), and hydrothermally altered rock colored gray with clay, white with zeolites and sulfates, yellow with sulfur, and brown with iron. Below this geothermal field lies a hydrothermal fracture network, with mineral precipitation doubtless taking place in many of the fractures. Bumpass Hell geothermal field, Lassen National Park, California, USA.

Figure 19.4 Eroded, dormant part of a geothermal field showing patches of dark, fractured, relatively fresh rhyolite, gray clay, white sulfates and zeolites, and brown iron staining. Notice how the alteration is not uniform. Highly altered fractures and sub-vertical regions occur between masses of less-altered rock. Behind the ponds, near the center of the image, is a young, fresh rhyolite lava flow, behind which is an active geothermal field that can be seen releasing steam. Landmannalaugar, Iceland.

Figure 19.5 Hydrothermally altered and mineralized fracture zone in quartzite. More or less fresh quartzite is to the left, and the alteration zone is to the right, with rusty staining from the weathering of sulfides. The deep pit just above the hood of the vehicle is where a mass of sulfide has weathered out. Wellesley Island, St. Lawrence River, New York, USA.

Figure 19.6 Terminated quartz crystals partially filling a vein in tonalitic gneiss. The fact that the quartz occurs as terminated crystals indicates that the fracture was never completely filled with quartz, and was never forced closed by lithostatic pressure. This means that the vein was deposited at a relatively shallow depth where the rocks were strong enough to hold the walls apart until the vein reached the surface. These crystals, especially near their terminations, are quite transparent, but typically vein quartz is loaded with fluid inclusions which make it look white. Lensvik, Trondheimsfjord, Sør Trøndelag, Norway.

Figure 19.7 Quartz – calcite vein in chlorite grade phyllite. This recently exposed vein has terminated quartz and calcite crystals that grew into fluid-filled space. This relatively planar fracture cuts across the rock foliation and cleavage, and so obviously post-dates ductile deformation of the rock. Springfield, Vermont, USA.

Figure 19.8 Needle-shaped, pink zeolite crystals in a narrow fracture in amphibolite. The lack of free-standing, terminated crystals indicates that this fracture was completely filled. The presence of zeolite suggests zeolite facies metamorphic conditions at the time of vein filling, though that was long after regional prograde metamorphism to epidote amphibolite facies in this region. Stangvik, near Surnadal, Møre og Romsdal, Norway.

Figure 19.9 This rock is garnet pyroxenite, broken from a nearby layer that is hosted in garnet peridotite (brown rock to the left). The garnet pyroxenite has been fractured, and the fractures are filled with serpentine. The garnet and diopside of this rock are unlikely to have been the source of the chemical components needed to make serpentine. The veins are interpreted to have formed from hot aqueous fluids that passed through the garnet peridotite, dissolving some olivine as it went. The fluids then flowed into joints in the garnet pyroxenite, where serpentine precipitated. The width of the field is about 20 cm. Ugelvik, Otrøy, Møre og Romsdal, Norway.

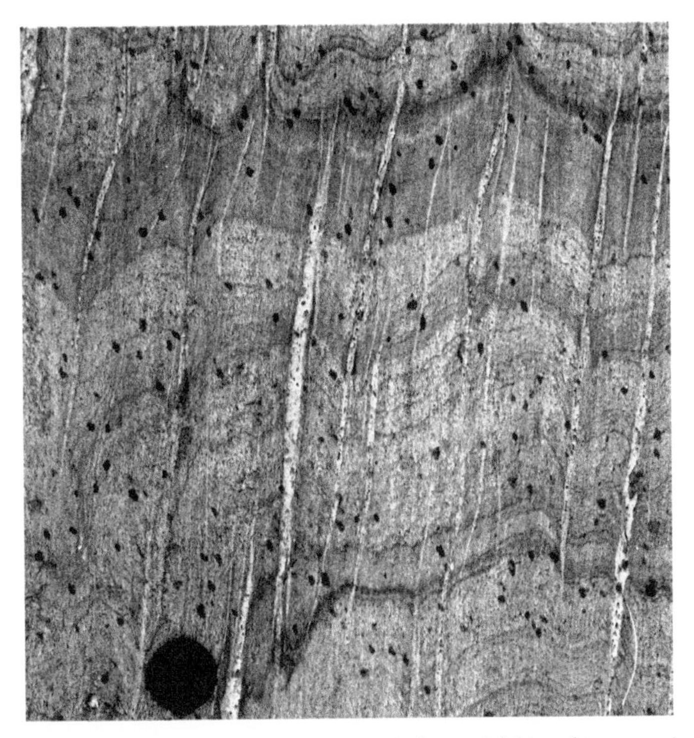

Figure 19.10 The host rock to the white quartz veins is layered feldspathic quartzite. The layering is related to what was originally sedimentary bedding, now probably highly transposed (Fig. 15.1A, B). The thin, white quartz veins were emplaced parallel to the axial planar cleavage of large-scale folds that are related to the small-scale folds visible here. These quartz veins seem to be in approximately their original shapes, and not boudinaged. Eastern Trollheimen, Oppland, Norway.

Figure 19.11 Chaotic-looking, white quartz veins cutting chlorite – biotite schist metamorphosed to biotite grade conditions. The dark, horizontal rock body cutting through the middle of the image is a metamorphosed basaltic dike. The dike is clearly cut by the quartz veins, and so predates them. The quartz veins seem to be oriented in many different directions. This is partly the result of these rocks being inhomogeneous, causing the directions of deformation (strain field) to have varied across the outcrop. It is also likely that the veins did not all form at the same time. Central Maine, USA.

Figure 19.12 Deformed white quartz veins in garnet – sillimanite – muscovite schist, with minor impure quartzite layers (yellow arrows). Veins like these may have originated at lower metamorphic grade, like those in Figure 19.11. Note the boudin neck to the lower left, between the red arrows. The boudin is subtle, but can be traced by following the quartzite layers. In this rock, some of the sillimanite-rich schists were less ductile than the quartzites and quartz veins. The light-colored rock in the lower-right corner is a large quartz vein. Jaffrey, New Hampshire, USA.

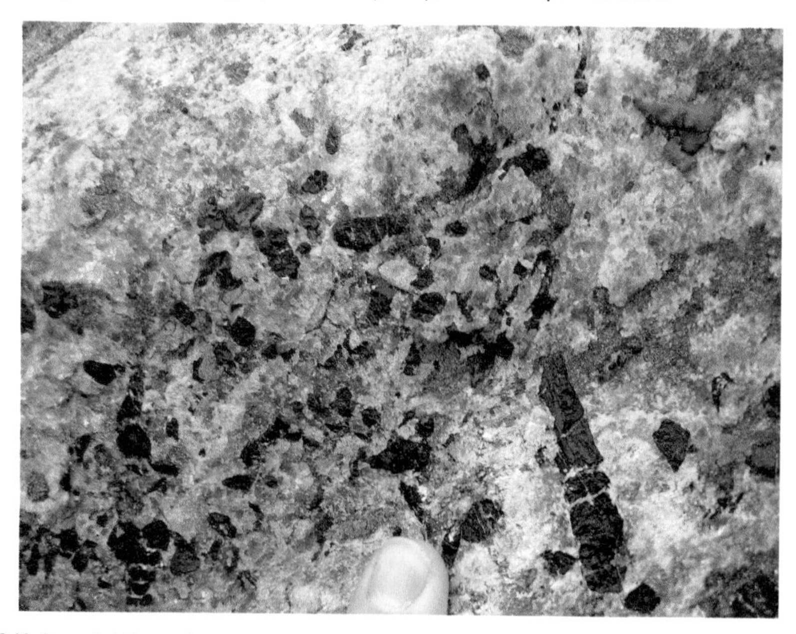

Figure 19.13 Post-folding, planar quartz – tourmaline vein, with coarsely crystalline (but not terminated) quartz and subhedral to euhedral tourmaline crystals. Some of the tourmaline crystals are fractured, like the big one to the lower right, with the fractures filled with quartz. Fracturing of tourmaline followed by quartz infilling probably took place during repeated fracturing of the vein as it successively opened and filled with precipitated minerals. The contact between the host schist and the vein is visible in the extreme upper left corner. Jaffrey, New Hampshire, USA.

Figure 19.14 Diopside – hornblende – quartz vein cutting fine-grained, diopside-bearing calc-silicate rock. The vein center is solid quartz, and the margins are mostly diopside and hornblende. This suggests that the mineral-depositing fluid changed composition as the fracture was filled. Wellesley Island, Adirondack Lowlands, New York, USA.

Figure 19.15 Coarse, almost pegmatitic quartz – feldspar – epidote – garnet vein cutting amphibolite. The mineral assemblage suggests that the vein formed under epidote amphibolite facies conditions, possibly at a slightly lower metamorphic grade than the enclosing epidote-free amphibolite. The metamorphic grade was, however, too low for the rocks to have melted (there is no evidence of melting), so this is indeed a vein rather than a pegmatite derived from silicate liquid. Lepsøy, Nordøyane, Møre og Romsdal, Norway.

Figure 19.16 This vein cuts tonalitic gneiss (G), some of which is visible to the left and lower right. The margin of the vein against gneiss is made mostly of prismatic white scapolite crystals and granular quartz (red arrow). The vein center is made of white, compact vein quartz (Q). The change in mineralogy indicates that the mineral-precipitating fluid changed composition during vein filling. The width of the field is about 40 cm. Lepsøy, Nordøyane, Møre og Romsdal, Norway.

Figure 19.17 Boudinaged sulfidic hydrothermal vein containing the mineral assemblage pyrite – magnetite – chalcopyrite – chlorite – epidote. This vein remnant was probably deposited in a submarine hydrothermal system, such as those that produced the massive sulfide ore deposits at nearby Løkken. Esvikneset, outer Trondheimsfjord, Sør Trøndelag, Norway.

Chapter 20

Metasomatism

Metasomatism involves a recognizable compositional change in a volume of rock caused by the gain, exchange, or loss of chemical components. Metasomatism can result in chemical or isotopic changes, which are commonly accompanied by mineralogical and textural changes. The movement of chemical components or isotopes at various scales during metamorphism is a routinely important process, as the development of porphyroblasts (Chapter 17) and veins (Chapter 19) should demonstrate. The results of metasomatism can be dramatic, transforming the parent into a mineralogically and texturally very different rock. The changes may also be subtle, as in the change of feldspar composition, trace element concentrations, or isotope ratios, with potentially no obvious change in rock textures or mineralogy.

In terms of the relative speed of transport of chemical components, fluid flowing along a fracture network is typically the fastest, followed by fluid flow along grain boundaries. Diffusion through intergranular fluid is usually slower than fluid flow, and diffusion in the solid state, along fluid-free grain boundaries and especially through crystal lattices, is generally the slowest. Movement in a flowing fluid is called advective or infiltration transport, and movement by diffusion is, amazingly enough, called diffusive transport.

Metasomatic changes are driven by different volumes of rock that are out of equilibrium with one another, being put into chemical communication by diffusion or fluid flow. Figure 20.1 shows an example in which blocks of ultramafic rock, composed mostly of antigorite and magnesite, were enclosed in granodiorite gneiss. The mineral assemblages in the ultramafic rock and the host gneiss were not in equilibrium with one another. During prograde metamorphism the two were put into chemical communication via a grain boundary fluid phase, through which chemical components could diffuse.

The diffusing chemical components allowed reactions to take place that transformed the original antigorite – magnesite mineral assemblage in the ultramafic rock interior to new assemblages. The reactions progressively consumed some chemical components from the fluid, but released others, and thus changed the fluid composition between the block center and the exterior gneiss. This resulted in concentric layers, each having a different mineral assemblage and local fluid composition. Different fluid compositions in different layers meant that there were concentration (actually activity, effective concentration) gradients, down which the chemical components diffused. The reactions continued to run as more of the chemical components diffused through the fluid toward places where reactions were consuming them. If the gneiss and ultramafic rocks had been in equilibrium originally, diffusion would have occurred anyway but there would have been no particular change to the rock mineralogy, composition, or texture.

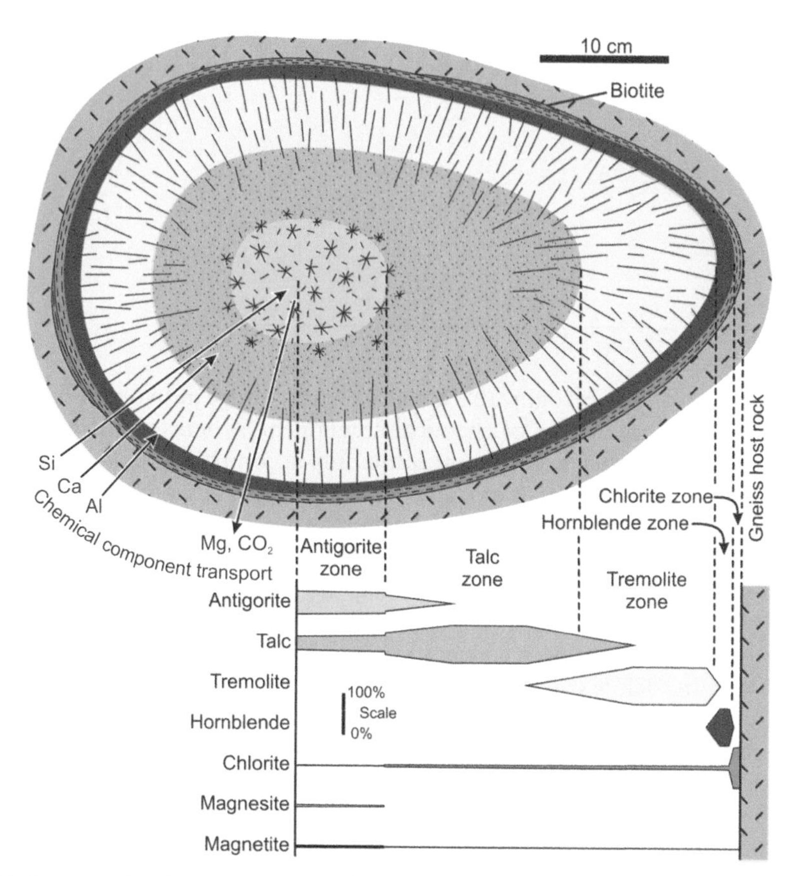

Figure 20.1 An example of metasomatism, where an ultramafic block, surrounded by granodiorite gneiss, was partially transformed by diffusive transport of chemical components through a grain boundary fluid. The gneiss host rock supplied Al, Ca, and Si, which diffused different distances into the ultramafic block. The block interior supplied Mg and CO_2, which diffused outward. In this case it was difficult to locate the original position of the ultramafic rock-gneiss contact, but it was probably at about the hornblende zone-chlorite zone boundary. These zoned ultramafic blocks were interpreted as having originally been dunite blocks that were serpentinized during low grade metamorphism. Later, metamorphism at amphibolite facies conditions produced the layering observed here. Adapted from Fowler et al. (1981).

Many processes can put rocks that are out of equilibrium with one another into chemical communication. Here are some examples:

1. A fracture allows aqueous, sulfide-bearing fluid from somewhere else to infiltrate an anhydrous mafic rock. The fluid infiltrates along grain boundaries, converting the anhydrous assemblage into an amphibole- and sulfide-bearing assemblage.
2. A fracture at the margin of a crystallizing tourmaline-bearing granite carries boron-rich fluid outward into pelitic schist. The fluid infiltrates from the fracture into the schist along grain boundaries, reacting with the rock to form tourmaline from boron, H_2O, minerals present in the host rock, and possibly other components from the fluid.

3. A hot basaltic pluton intrudes chlorite grade phyllite. Contact metamorphism causes the hot phyllite to react to form anhydrous phases. The released fluid dissolves some of the rock, and migrates away to cooler regions, reacting with the rock as it moves.
4. A brittle fault puts cold dolomitic marble against cold granite. During later metamorphism, silica from the granite diffuses into the marble to produce Ca-Mg silicates such as tremolite and diopside, and Ca and Mg diffuses out of the marble and into the granite to produce hornblende, titanite, and Ca-rich plagioclase.

Evidence of metasomatism at thin section scale implies either slow material transport (e.g., by solid-state diffusion), a short period of time for the reactions to have taken place, or a small amount of available external material. Metasomatism on an outcrop scale or larger implies either rapid material transport (e.g., fluid flow), or a long period of time and large amounts of available external material.

Why does the explanation above, and elsewhere in this book, keep using the word "fluid" instead of, say, 'water'? First, H_2O may not have been the major volatile component in the fluid. Where fluid is associated with carbonates, it may be rich in or dominated by CO_2. Other volatile components like methane, nitrogen, noble gasses, or hydrogen sulfide may also be important in some places and times. Second, to a petrologist the word 'water' refers to the dense liquid, like ice refers to the cold solid and water vapor refers to the same compound as a gas. Ice is not a fluid, and seems unlikely to be important during rock metamorphism in any case, but water and water vapor certainly are fluids. In addition, above its critical point (374°C, 218 bars for pure water), H_2O is neither water liquid nor water vapor, but a supercritical fluid that can transition from a dense, low volume fluid to a low-density, high volume fluid with no phase change (no distinct transition, no meniscus). The same is true of other supercritical fluids. Fluid mixtures, with and without dissolved solids, can have additional complications such as two immiscible fluids (e.g., Kaszuba et al., 2006). The point is that the generic word 'fluid' lets us avoid having to worry about what the phase actually was, and to be comfortably vague about its composition.

The thin section images in Figure 20.2 show five examples of metasomatism. Figure 20.2A shows the transformation of a pyroxenite into amphibolite by movement of H_2O, SiO_2, Ca, and CO_2. Figure 20.2B shows the metasomatic replacement of hornblende by riebeckite, caused by Na-rich hydrothermal fluid. Figure 20.2C and D are from a single thin section, showing concentric reaction rims around ilmenite (C) and olivine (D) that formed at granulite facies conditions from solid state diffusive transport. Figure 20.2E is a before picture, of a sillimanite – staurolite grade pelitic schist. Figure 20.2F is the after picture, where boron-rich fluid from a quartz vein has transformed the schist into a quartz – tourmaline – garnet – muscovite rock (see Fig. 20.5).

The field photos first show metasomatic effects surrounding fractures that once carried fluid (Figs. 20.3–20.7). Following that are metasomatic halos around pegmatites (Figs. 20.8, 20.9). Figures 20.10 and 20.11 show metasomatic zones between marble and adjacent silicate rock, and Figure 20.12 shows metasomatic effects of a sedimentary xenolith in gabbro. Figures 20.13 and 20.14 show the transition zone between olivine-bearing gabbro, almost undeformed but metamorphosed to granulite facies, and garnet amphibolite in which garnets up to 15 cm in diameter are visible. Lastly, Figure 20.15 shows concentric mineral shells that grew during granulite facies, solid-state diffusive transport of chemical components, like that in Figure 20.2D.

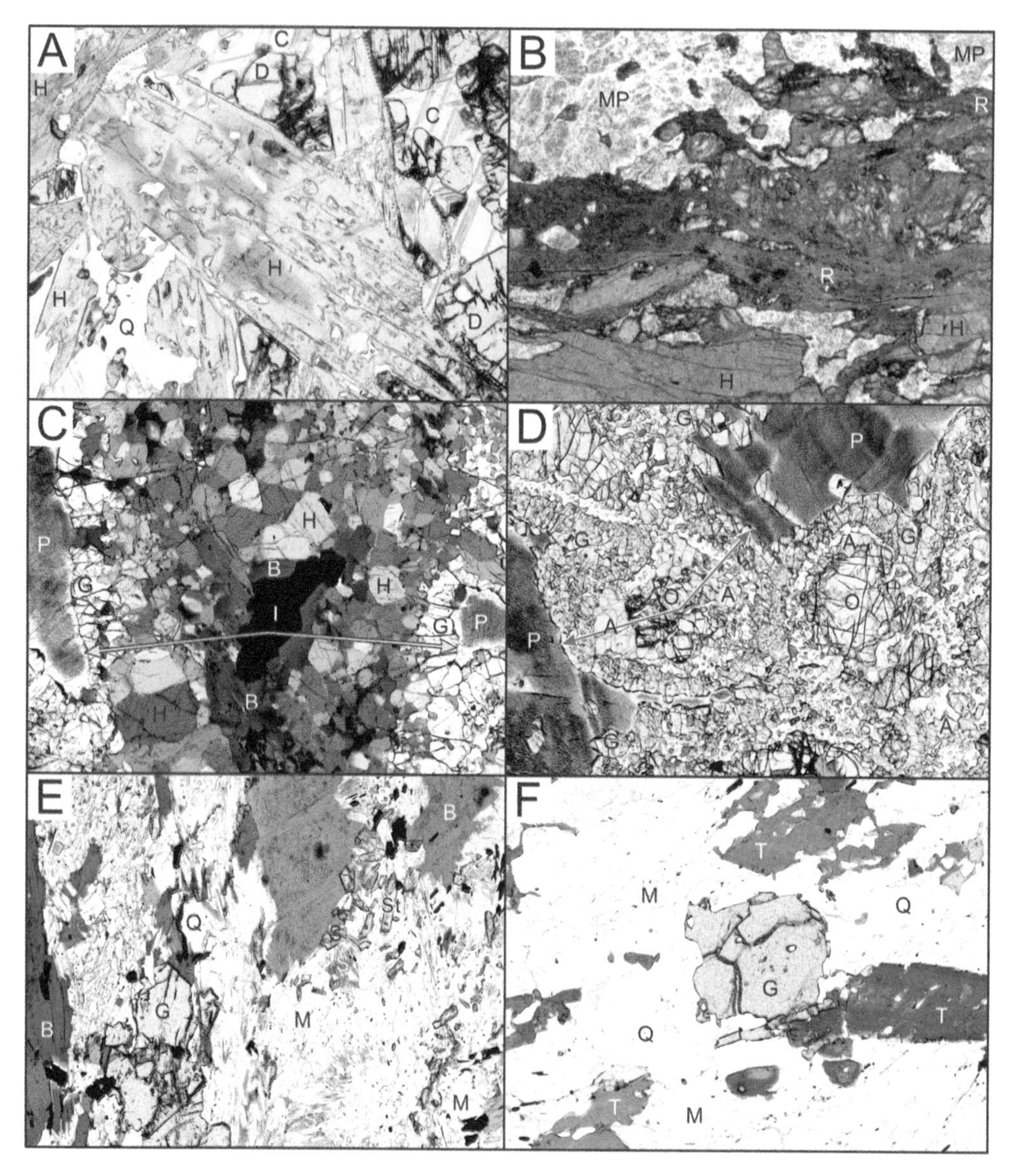

Figure 20.2 Photomicrographs showing examples of metasomatism at thin section scale. All images are in plane-polarized light, and have field widths of 4 mm. A) Metasomatic reaction zone between a calcite-bearing clinopyroxenite (right) and quartz-bearing hornblendite (upper left). The metasomatic reaction assemblage (mottled hornblende, quartz, diopside, calcite) is between the red dotted lines. In the reaction zone, the light-colored hornblende is interpreted to have directly replaced diopside, and the darker parts grew in interstitial regions. The hornblende is full of quartz inclusions because the transformation of diopside to hornblende produces excess quartz. To run the reactions, H_2O, Al, Fe, and Na were transported to the right, and CO_2 and Ca transported to the left. Northfield, Massachusetts, USA. B) Fractured granitic pegmatite that has been partially metasomatized by oxidizing, Na-rich fluid, replacing yellow-brown hornblende with blue riebeckite. The source of the fluid is interpreted to have been hydrothermal groundwaters derived from saline playa lakes that once existed in the Triassic-Jurassic Newark rift basin. Mendham, New Jersey, USA. C, D) These two images are of the same thin section, of an olivine gabbro that was metamorphosed to granulite facies conditions.

Figure 20.2 (Continued) In the absence of fluid, chemical components diffused in the solid state to produce 'corona' or 'moat' structures, visible here. In C, the mineral layering sequence from corona center to rim is ilmenite, biotite, hornblende, and garnet, with external plagioclase + spinel. Yellow arrows extend from ilmenite to the plagioclase – garnet contact. Fe and Ti diffused outward, and Si, Mg, K, Na, and Ca diffused inward. In D, the mineral layering sequence is olivine, augite, plagioclase, and garnet-plagioclase symplectite, with external plagioclase + spinel. Red arrows extend from olivine to the plagioclase-garnet contact. In this case, Mg diffused outward, and Si, Ca, Al, and Fe diffused inward. In both C and D the external plagioclase is dark because it contains abundant, minute green spinel inclusions that formed from excess Mg, Fe, and Al. North River, Adirondacks, New York, USA. It may be argued that C and D do not show metasomatism, but nonetheless chemical components were moving down activity gradients, forming mineral zones that otherwise would not have occurred. E) Pelitic schist having the assemblage quartz – sillimanite, – muscovite – biotite – staurolite – garnet (before metasomatism view). Jaffrey, New Hampshire, USA. F) A metasomatized version of E, taken from about one meter away in a metasomatic zone similar to that seen in Figure 20.5. In this zone, boron-bearing fluid infiltrated the surrounding schist. All of the biotite, sillimanite, and staurolite are gone in F, leaving quartz, garnet, muscovite, and lots of new tourmaline. Jaffrey, New Hampshire, USA. Abbreviations: A, augite; B, biotite; C, calcite; D, diopside; G, garnet; H, hornblende; I, ilmenite; hornblende; M, muscovite; MP, microcline perthite; O, olivine; P, plagioclase; Q, quartz; R, riebeckite; S, sillimanite; St, staurolite; T, tourmaline.

Figure 20.3 Metasomatized zone surrounding fractures in fine-grained pink granitic gneiss. Nearby there is a small pyrite deposit associated with intensely fractured and hydrothermally altered quartzite (Fig. 19.5). Small fractures extend from the intensely altered zone into the surrounding rocks. Fluids flowing through the fractures here infiltrated several millimeters into the gneiss, and reacted with it to form dark hornblende and diopside in the metasomatic halo. Wellesley Island, Adirondack Lowlands, New York, USA.

Figure 20.4 Relatively undeformed quartz vein in muscovite – biotite schist, emplaced parallel to the dominant rock cleavage. Discontinuous dark inclusions in the quartz vein (two indicated with a red arrow) are not xenoliths in the sense that they are fragments broken from the wall rock and suspended in magma. Because this is a vein, deposited by aqueous fluid, quartz was initially deposited in one fracture, filling it. Then, a second fracture and vein formed very close to the first, isolating a wall rock fragment. This process had to have happened several times to account for all the small, dark wall rock fragments in this quartz vein. The black margins of the quartz vein (yellow arrows) are mostly tourmaline, indicating that the quartz-depositing fluids contained boron, which metasomatized the wall rock within several centimeters of the vein margin. Big Thompson Canyon, Colorado, USA.

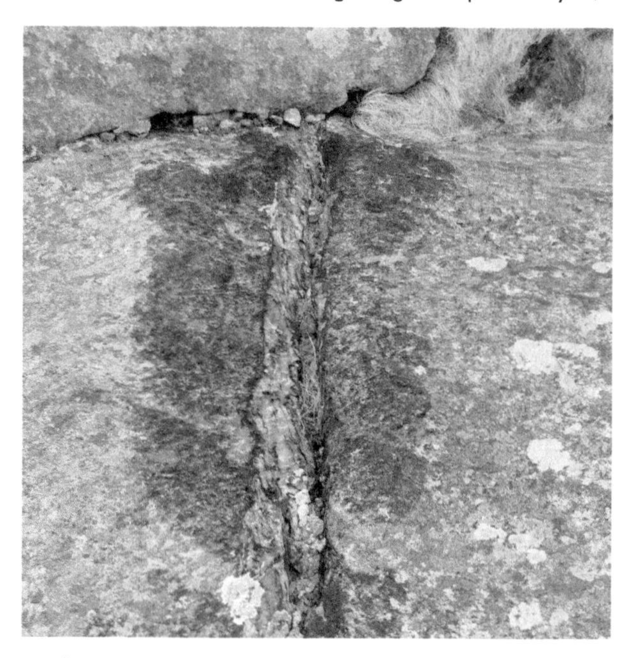

Figure 20.5 A 7 cm wide quartz – tourmaline vein, with 1–20 cm wide metasomatized zones to either side. The schist mineralogy is muscovite – biotite – garnet – sillimanite – staurolite – quartz. In the metasomatized zone the mineralogy is tourmaline – quartz – muscovite – garnet. Infiltration of boron-bearing fluid transformed essentially all sillimanite, staurolite, and biotite into tourmaline. The

Figure 20.5 (Continued) vein probably fractured and filled in several stages, as indicated by alternating, approximately wall-parallel quartz-rich and tourmaline-rich layers. These rocks are similar to those shown in Figures 19.2E, F and 20.2E, F. Jaffrey, New Hampshire, USA.

Figure 20.6 Olivine websterite (light-colored rock), that was cut by fractures and metasomatized by fluids to garnet – diopside – phlogopite websterite (dark rock). The parental olivine websterite last equilibrated at ultra-high pressure conditions (40 kbars, 800°C, contains microdiamonds), similar to the conditions when metasomatism occurred (Vrijmoed et al., 2006). This whole outcrop is an isolated block of mantle rock that is surrounded by granitic gneiss that also hosts bodies of eclogite. Figure 20.7 shows a close-up of one of the metasomatic veins. Svartberget, Bud, Møre og Romsdal, Norway.

Figure 20.7 Close-up of fresh surfaces of the outcrop in Figure 20.6, showing one of the metasomatic veins cutting mantle olivine websterite (OW). Greenish areas are diopside – garnet – phlogopite websterite, with bright reflections of sunlight from the cleavage surfaces of large phlogopite crystals. Garnet-rich zones are red. The width of the field is about 0.5 m. Svartberget, Bud, Møre og Romsdal, Norway.

Figure 20.8 White pegmatite dike cutting brown eclogite. The pegmatite intruded during amphibolite facies conditions, and supplied aqueous fluid that infiltrated the eclogite and transformed it to black amphibolite for a distance of 0.5–1.5 meters. Flemsøy, Nordøyane, Møre og Romsdal, Norway.

Figure 20.9 This image is of a layered gabbro (see Fig. 25.16) that has been metamorphosed to granulite facies conditions. Its originally almost anhydrous mineral assemblage was dominated by plagioclase, garnet, and pyroxenes, with minor hornblende and biotite. Brown rock with this assemblage is visible in the lower third of the image. The white pegmatite intruded under amphibolite facies conditions, supplying aqueous fluid that transformed the adjacent pyroxene granulite to dark-gray amphibolite. Rhythmic igneous modal cumulate layering is visible above and below the pegmatite dike, having survived initial granulite facies conditions and later transformation to amphibolite. Haramsøy, Nordøyane, Møre og Romsdal, Norway.

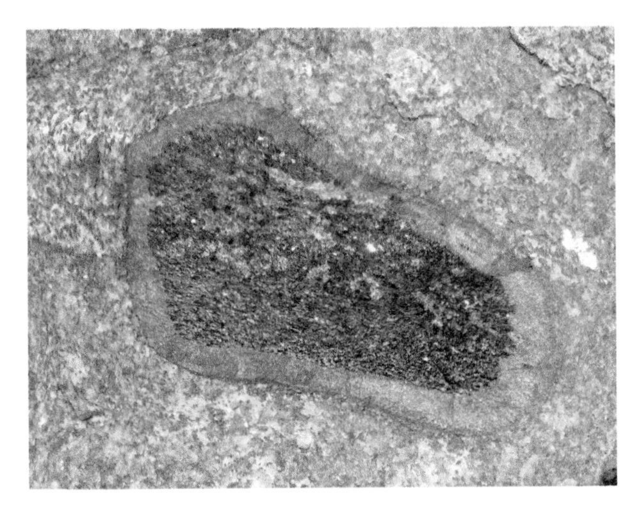

Figure 20.10 Xenolith-like block of amphibolite incorporated into coarse, gray-green calcite marble. The marble has the assemblage calcite – diopside – olivine – graphite. The interior of the block is dark-gray hornblende – plagioclase amphibolite. The intermediate, lighter-colored zone between the marble and amphibolite core is amphibolite that was metasomatically altered to calc-silicate rock, made mostly of diopside and anorthite. The width of the field is about 40 cm. Warrensburg, New York, USA.

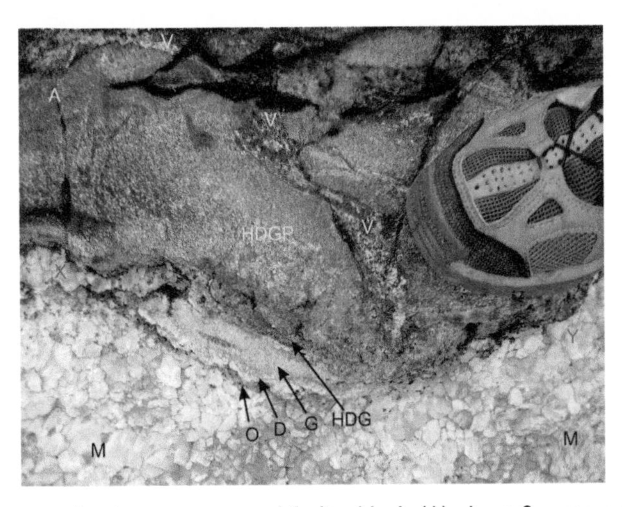

Figure 20.11 Contact region between an amphibolite block (A) about 2 meters across, in olivine-bearing marble (M). The marble contains the assemblage calcite – olivine – diopside – graphite. At the contact there are several mineralogically distinct layers: olivine-rich marble (O), diopside (D), grossular-andradite garnet (G), a layer of hornblende, diopside, and grossular-andradite garnet (HDG), and a broad, irregular zone of hornblende, diopside, grossular-andradite garnet, and plagioclase (HDGP). This represents the diffusion of different components across the amphibolite-marble boundary, principally Si and Al out from the amphibolite, and Ca in from the marble. The outer layers (O, D, G, HDG) are missing in whole or in part from numerous places around the amphibolite block. At the red X, marble occurs in place of layers D, G, and HDG, but layer O is apparently intact with calcite on either side. At the red Y, layers O, D, and G are missing, with marble in contact with HDG. This suggests a complex history involving early layer development, and later removal by different mechanisms, possibly by breaking (Y?) or selective dissolution along a fluid channelway (X?). V is a hornblende – diopside – anorthite vein that cuts through the amphibolite. The vein is truncated by the contact with marble, and so pre-dates incorporation of this amphibolite block into the marble. Warrensburg, New York, USA.

Figure 20.12 Contact metamorphosed sedimentary xenolith in dark-gray gabbro. The white to pink sedimentary rock (R) was thermally metamorphosed to a mixture of mostly plagioclase, rusty-weathering green spinel, and cordierite. The intervening greenish-gray rock (white arrow) has been metasomatized along bedding planes and crosscutting fractures by components from the gabbro. The metasomatized rock has an assemblage principally of plagioclase and diopside, indicating introduction of at least Ca to the metasomatized zone. Between the thermally metamorphosed rock and greenish-gray metasomatized zones, there are two thin intervening metasomatic layers: white (red arrow, probably plagioclase-rich), and black (yellow arrow, probably magnetite-bearing). These layers are early parts of the three-layer metasomatic sequence (black layer, white layer, greenish-gray rock). Skaergaard intrusion, east Greenland.

Figure 20.13 Quarry face at an abandoned garnet mine. The garnet amphibolite ore rock is to the left of the boundary indicated by the yellow arrows. The precursor rock, to the right, is layered olivine gabbro that was metamorphosed to granulite facies conditions. The transition zone between the two is about 2 m wide, over which metamorphic garnets less than 0.2 mm in diameter in the gabbro gradually decrease in number, and increase in size, to greater than 10 cm in diameter. Cumulate layering, indicated by red arrows, extends from the gabbro into the garnet amphibolite, indicating that the precursor rock transformed directly to garnet amphibolite with little deformation. The garnet amphibolite formed during infiltration metasomatism by an aqueous fluid at upper amphibolite facies conditions. The major element compositions of the gabbro and garnet amphibolite are nearly identical, except of course for H_2O. However, some trace elements in the garnet amphibolite are dramatically enriched or depleted, and inhomogeneous, compared to the gabbro (Hollocher, 2008; Morgan and Hollocher, 2011), indicating channelized fluid flow. North River, New York, USA.

Figure 20.14 Close-up of part of the gabbro-garnet amphibolite transition zone, like that in Figure 20.13, but in a different part of the mine. To the far left is brownish olivine gabbro, metamorphosed to granulite facies conditions, having garnets less than 0.2 mm in diameter. To the right is garnet amphibolite having garnets up to 2 cm in diameter. The width of the field of view is about 1 m. Gore Mountain, North River, New York, USA.

Figure 20.15 Image of a freshly broken rock surface of a corona gabbro, like that seen in Figures 20.2C and D, 20.13, and 20.14. This one, however, has coronas with the sequence: orthopyroxene (light-green, green arrow), diopside (white, white arrow), and garnet (pink, pink arrow). The olivine in corona centers was, in this case, completely consumed during granulite facies metamorphism. The dark greenish-gray minerals are blocky, original igneous plagioclase crystals (P) that are full of minute spinel inclusions. The field width is about 8 mm. North Hudson, Adirondacks, New York, USA.

Chapter 21

Partial melting and migmatites

"The resistance offered by limestones against granitization is very remarkable. Even in the midst of a migmatite area, where all siliceous rocks have been thoroughly mixed or assimilated with the granite magma, the limestones are generally quite free from granitic injections, and are intersected only by rectilinear dikes." Eskola (1922).

Migmatites have a long and controversial history, partly from their diverse appearance and partly from the many ways in which they have been interpreted to form. Migmatites are outcrop-scale rock assemblages that are mixtures of apparently metamorphic parts, (commonly schists, gneisses, or amphibolites), and igneous parts (commonly medium-grained to pegmatitic granite, granodiorite, or tonalite). The igneous-looking parts can occur in a variety of forms, including dikes, sills, parallel layers, and irregular patches, with sharp or diffuse margins. The great diversity of migmatite appearance has led to an impressively diverse terminology (e.g., Robertston, 1999), most of which will be avoided here. Generally, the igneous-looking parts are coarser-grained, richer in quartz and possibly feldspar, and lighter-colored than the rest, and are referred to as leucosomes. The metamorphic-looking rock from which the partial melt may (or may not) have been extracted is called the mesosome. In some cases, dark selvages, or melanosomes, separate the leucosome from the mesosome.

Partial melting represents the broad boundary between metamorphic and igneous realms. At moderate to high-grade metamorphic conditions (middle amphibolite to granulite facies) many rocks can undergo partial melting. The melt initially coats grain boundaries and, if the melt does not separate from the rock, it will eventually crystallize *in situ* back onto adjacent unmelted grains. This may leave behind no distinctive textures to indicate that melting ever took place. Commonly, however, the grain boundary melt separates from the rock by segregation into extensional fractures, where the melt prevents fracture closure because it is at or near lithostatic pressure. The residual rock, plus crystallized melt that traveled distances of perhaps centimeters or meters, is the migmatite. Melting can be initiated by rising temperature, decreasing lithostatic pressure, H_2O-releasing dehydration reactions, or an influx of fluid from an outside source.

Melting is commonly related to one or more prograde metamorphic dehydration reactions involving an abundant mineral, such as muscovite in pelitic rocks or amphiboles in amphibolites. The H_2O released by a dehydration reaction acts as a flux to stabilize hydrous silicate liquid. Small amounts of melting produce what are essentially minimum-temperature melts for the system, with melt initially coating grain boundaries (Fig. 21.1Aa, Ab). Melting under these circumstances is eutectic-like, with the same mineral assemblage progressively producing melt of almost the same composition over a small temperature interval (this is a basic concept of igneous petrology).

Eventually, one phase important to the melting reaction may be exhausted (quartz in Fig. 21.1Ab). If the melt escapes at this point, a restite (residual solid) assemblage is left behind with one phase absent (Fig. 21.1Ac). This new, simpler assemblage has a substantially higher melting temperature than the original, so moderate increases in temperature will not produce more melt (Fig. 21.1Ad). If instead the melt does not separate after exhaustion of a phase (Fig. 21.1Ae), at least temporarily, melting will continue as the temperature rises (Fig. 21.1Af).

Figure 21.1Ba shows a somewhat different system. Silicate liquid produced during melting initially coats grain boundaries (Figs. 21.1Bb), and can segregate into opening fractures the same way fluid can (Fig. 21.1Bc). Silicate liquids produced by partial melting are generally more felsic (richer in Si, K, Na, poorer in Mg, Fe, Ca) than the parental rock, and especially the residual restite. The liquids are also richer in H_2O. Because of the composition difference, the crystallized liquid will typically be richer in quartz and possibly feldspar than the parental rock, and poorer in mafic minerals (Fig. 21.1Bd, H_2O-rich fluid escapes). However, because liquid and restite were in equilibrium before segregation, both should have identical mineralogy (Fig. 21.1Bd) unless: 1) melting proceeded to the exhaustion of a phase, and 2) melt was 100% extracted to the segregations (Fig. 21.1Cd). If the melt crystallizes and dissolved fluids do not escape, or if fluids enter from elsewhere, the restite and leucosome may be retrograded back to the pre-melting mineralogy, though textures indicating melting and melt separation will remain (Fig. 21.Ae).

The origin of melanosomes is more difficult to interpret. If, in a partially melted rock (Figs. 21.1Ca, Cb), all melt segregates into leucosomes (Fig. 21.1Cc), the result would be a two-part rock (Fig. 21.1Cd): the leucosome and the restite (region X). If instead, melt segregated from only that part of the rock nearest the opening fracture (Fig. 21.1Ce), the result would be a three-part rock (Fig. 21.1Cf): a mesosome with the original rock composition (region Y, because the melt stayed), the leucosome (where melt adjacent to the fracture went), and a melanosome (region Z, like region X in Fig. 21.1Cd).

This concept of mafic selvage (melanosome) development may be correct in some cases, but there are other possibilities. 1) The mafic selvage may have been precipitated by the melt as it crystallized, or perhaps as it flowed through the fracture (Figs. 21.11, 21.12, 21.16?). 2) Melt from elsewhere, flowing through the fracture, may have dissolved out the more felsic wall rock components, leaving behind a mafic selvage. 3) In some cases, particularly where the melanosomes are discontinuous, anomalously thick, or chemically or mineralogically distinct from adjacent rocks, the melt may simply have entered a fracture that developed along a weak mafic layer. In this instance, the melanosome has no genetic relationship to the melt or melt separation process (Fig. 21.15?).

The discussion above describes migmatites that are derived from the segregation (Figs. 21.3–21.10), and short-range flow (Figs. 21.11–21.15), of partial melts that formed during metamorphism. Is this how all migmatites formed? Almost certainly not. Locally, melts may have been introduced originally as dikes or sills, from a more extensive, nearby magmatic source. In some cases such a relationship can be demonstrated in the field, but in others deformation, recrystallization, and surface cover may obscure it. If so, determining migmatite origin may be difficult (e.g., Lancaster et al., 2009). Alternatively, migmatites may result from deformation of pre-metamorphic inhomogeneities, such as interlayered sediments and volcanics, or dikes cutting a host rock. During metamorphism, one layer type may recrystallize, or partially melt, giving the rock a migmatitic appearance without partially melting the apparent mesosome, and without melt segregation. Lastly, some migmatites may have leucosomes that precipitated from hot fluids migrating through extensional fractures, rather than by segregation of partial melts.

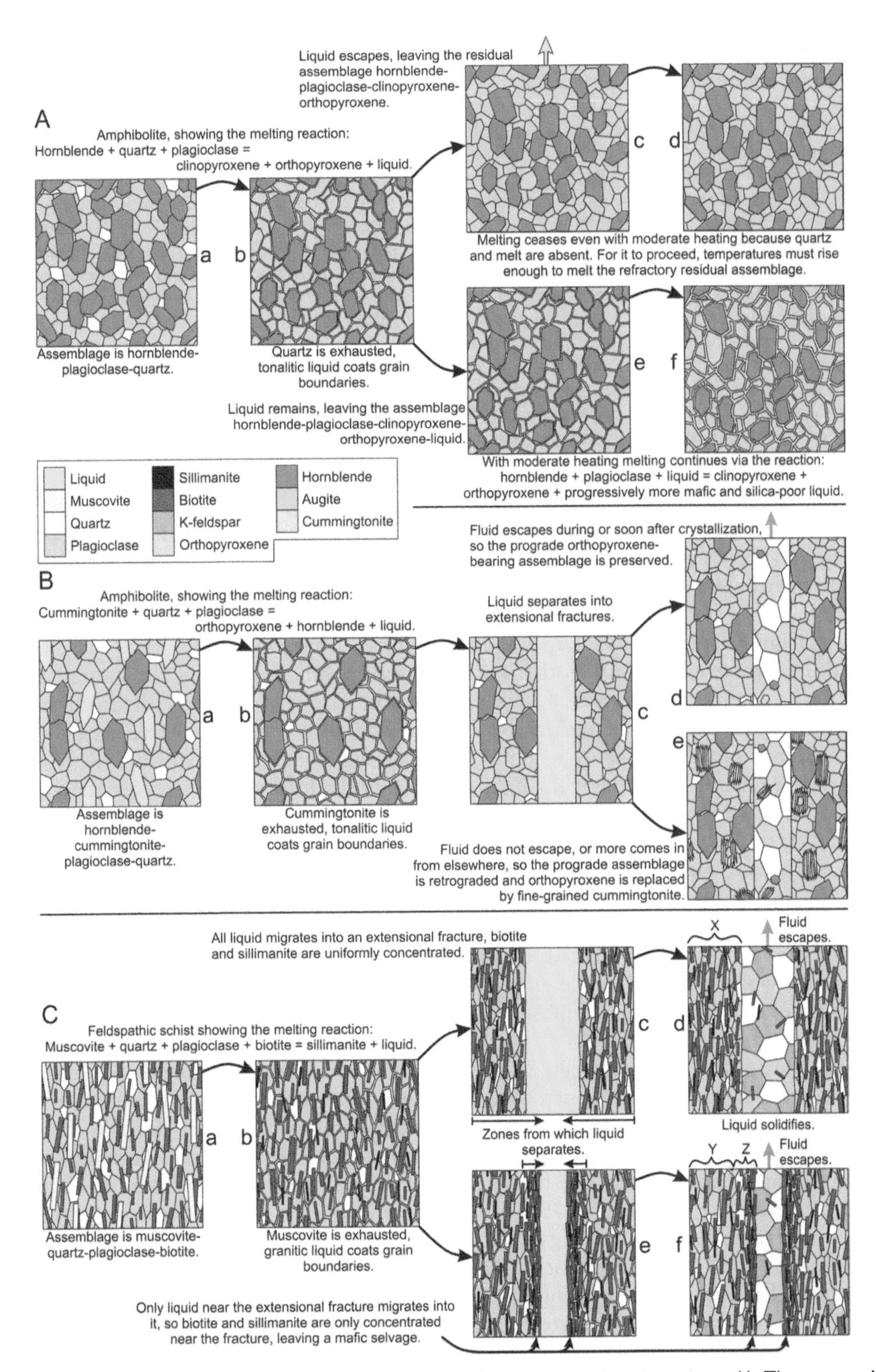

Figure 21.1 Illustration of melting reactions and melt separation in migmatites. A) The parental rock (Aa) undergoes partial melting to the exhaustion of quartz (Ab). If the liquid escapes (Ac),

Figure 21.1 (Continued) melting ceases (Ad) because the quartz-free assemblage is more refractory than the original. For melting to proceed, temperatures must rise to the melting temperature of the residual plagioclase – hornblende assemblage. Alternatively, if the liquid does not escape after quartz exhaustion (Ae), melting can progress as temperatures continue to rise (Af). B) The parental rock (Ba) undergoes partial melting (Bb) to the exhaustion of cummingtonite, resulting in an orthopyroxene-bearing mineral assemblage and grain boundary liquid. The liquid separates into an extensional fracture (Bc), which eventually crystallizes to a relatively coarse-grained, tonalitic rock that is more felsic than the solid restite left behind. If the fluid escapes (Bd), the prograde assemblage remains intact. If the fluid remains, or is later supplied from an outside source, the prograde assemblage in both the host rock and leucosome is retrograded (Be). C) Example of one way selvages can form at the margins of leucosomes. A muscovite schist (Ca) undergoes partial melting to the exhaustion of muscovite, changing the assemblage and producing grain boundary liquid (Cb). The liquid completely separates from the rock into an opening fracture (Cc), eventually crystallizing to a granitic leucosome. All of the residual rock (region X) is restite, the solids left over from melting, and no melanosome develops. In a different circumstance, only the liquid nearest the fracture successfully migrates into it (Ce), enriching only region Z, adjacent to the fracture, in residual solids (exaggerated biotite content in Ce and Cf to make the point clear). After crystallization (Cf), the rock distant from the fracture (region Y) has the same chemical composition as the original rock, and a mafic selvage (region Z, melanosome) is left next to the granitic leucosome.

The impression from Chapter 19 may be that it is easy to tell the difference between veins and melt segregations, but Figure 19.15 indicates that this is not always true. That figure shows a vein dominated by intergrown quartz and feldspar, a texture that might be interpreted as igneous (i.e., a crystallized dike). However, the odd assemblage, particularly the large amount of epidote, the connection of the thick vein to thin, epidote-rich fractures that cut the host amphibolite (top of the figure), and the lack of other indications of melting in the host rocks, suggests a hydrothermal origin. Nonetheless, distinguishing between feldspar-rich veins and partial melt segregations may not always be easy. Finding crystals that coarsen inward from the leucosome margins, or mineralogically distinct layers in the leucosomes (e.g., quartz cores, Figs. 19.14, 19.16), may point toward deposition by flowing fluid, rather than silicate liquid crystallization.

Lastly, rock melting and melt migration during metamorphism, at the extreme, can supply the huge volumes of magma needed for the many observed felsic and intermediate extrusive and plutonic rock bodies. How and where sufficient volumes of magma were produced, and how they migrated, is a much-discussed topic (e.g., Brown et al., 2011). The examples of migmatites shown here, however, probably involved only small volumes of melt that didn't migrate far before crystallizing from cooling or fluid escape.

The thin section photos (Fig. 21.2) show three examples of host/migmatite rock pairs derived from pelitic schist, tonalitic gneiss, and amphibolite. In each case the leucosome is interpreted to have been segregated partial melt that formed during metamorphism, though it can't be proven that the melt in the thin section came from exactly the adjacent rock. The field photos start with migmatites that have discontinuous leucosomes that probably did not involve partial melt movement by more than a several centimeters (Figs. 21.3–21.10). This sequence includes mafic rocks (Figs. 21.3–21.6), granitoid rocks (Figs. 21.7, 21.8), and pelitic schist (Figs. 21.9, 21.10). The next set shows migmatites where the leucosome material has apparently migrated on a scale of meters, presumably as silicate liquid (Figs. 21.11–21.15). Figures 21.15 and 21.16 show melanosome selvages, and Figure 21.17 shows two generations of migmatite, one before and one after folding.

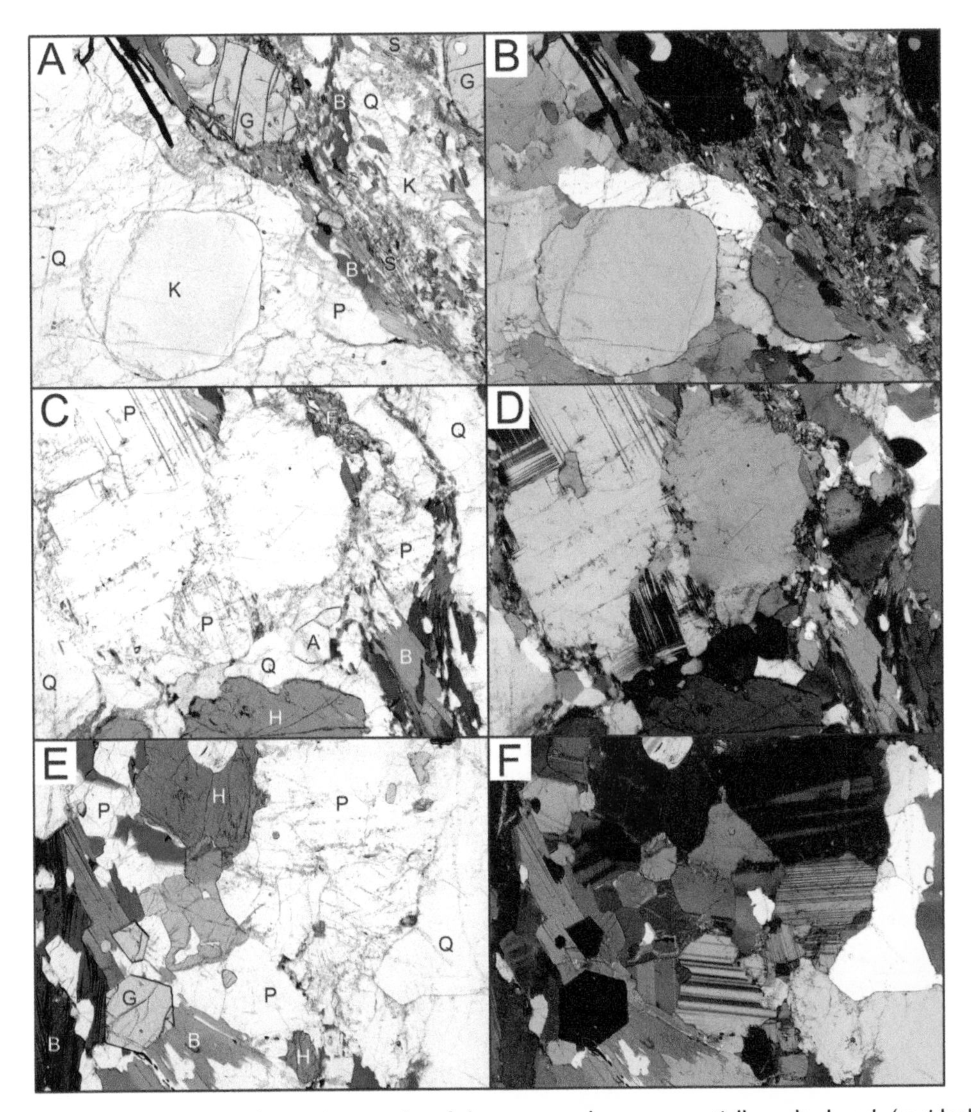

Figure 21.2 Thin section photomicrographs of the contacts between partially melted rock (residual mesosome) and the extracted melt (leucosome). Images on the left are in plane-polarized light, those on the right are the same fields in cross-polarized light. All field widths are 4 mm. A, B) Pelitic schist at sillimanite – K-feldspar grade, with the leucosome in the lower left half. The schist assemblage is garnet – biotite – K-feldspar – quartz – plagioclase – sillimanite. Like other schists in the area, the pegmatitic leucosome is presumed to have formed during muscovite dehydration. The fluid escaped during crystallization and the rock remained dry, as indicated by the lack of retrograde muscovite. The K-feldspar in this rock is microperthite, causing the apparent cloudiness of the one large K-feldspar crystal (K). Ware, Massachusetts, USA. C, D) Hornblende – biotite tonalitic gneiss (approximately the right quarter) adjacent to a more felsic quartz – plagioclase – hornblende – biotite leucosome (left). The leucosome has euhedral hornblende crystals visible on the outcrop surface. Quabbin Reservoir, Massachusetts, USA. E, F) Biotite – garnet amphibolite with the assemblage hornblende – plagioclase – biotite – garnet – quartz (left half), in contact with a tonalitic leucosome that has the same assemblage but with more plagioclase and much more quartz, which is almost absent in the host rock. This rock melted during upper amphibolite facies conditions. Ware, Massachusetts, USA. Abbreviations: A, apatite; B, biotite; E, epidote (retrograde in C, D); G, garnet; H, hornblende; K, K-feldspar; P, plagioclase; Q, quartz; S, sillimanite.

Figure 21.3 This amphibolite was metamorphosed to upper amphibolite facies conditions. Partial melting was apparently induced by cummingtonite dehydration, with the released H_2O stabilizing tonalitic liquid. The liquid migrated into irregular, discontinuous extensional fractures that opened during deformation, producing tonalitic leucosomes with the assemblage plagioclase – quartz – biotite – hornblende. The leucosomes also have small amounts of fine-grained, twinned cummingtonite, apparently from the retrograde conversion of prograde orthopyroxene (Fig. 21.1Ae). The host amphibolite has the same assemblage and textures, though with almost no quartz. Figure 21.4 is a close-up of the leucosome below and to the right of the hammer point. Ware, Massachusetts, USA.

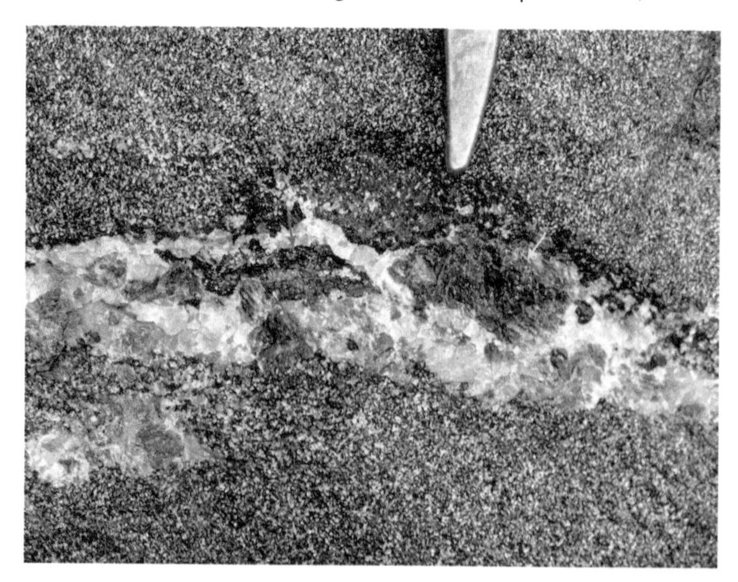

Figure 21.4 Close-up of one of the tonalite-filled fractures shown in Figure 21.3, with coarse, dark-gray, unrecrystallized plagioclase (yellow arrow points to the largest) and recrystallized white plagioclase and quartz. Minor mafic minerals in the tonalite are hornblende, biotite, and fine-grained cummingtonite that is interpreted to be retrograde after orthopyroxene. The host amphibolite has an identical assemblage, except that quartz has partitioned almost entirely into the leucosomes. The gray color of the plagioclase is caused by minute exsolved Fe-Ti oxide rods and plates, which is characteristic of some igneous plagioclase. The red arrow points to a rock fragment enclosed by the leucosome. Ware, Massachusetts, USA.

Figure 21.5 Fine-grained amphibolite hosting a coarse-grained tonalitic leucosome that is interpreted to be a segregated partial melt from the amphibolite. The amphibolite assemblage is plagioclase – hornblende – garnet – cummingtonite – biotite, with traces of quartz. The leucosome assemblage is the same, but with vastly more quartz, somewhat more plagioclase and cummingtonite, and less hornblende, biotite, and garnet. Hornblende is black, and the large cummingtonite crystals in the leucosome are grayish-green. Ware, Massachusetts, USA.

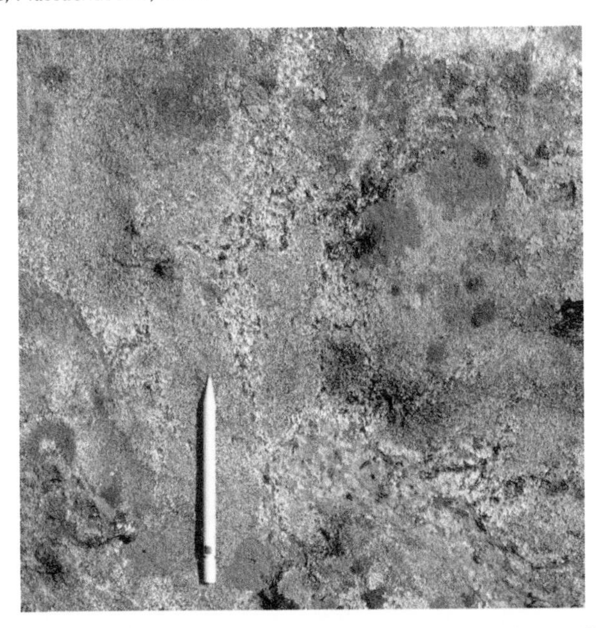

Figure 21.6 This is a dioritic rock metamorphosed to granulite facies conditions. Both the fine-grained rock and coarse-grained leucosomes have the same assemblage: plagioclase – augite – orthopyroxene – quartz – biotite. The leucosomes contain almost all of the quartz and somewhat more plagioclase, but less augite, orthopyroxene, and biotite than the fine-grained surrounding rock. The rock is compositionally similar to some amphibolites found at lower grade. In this case melting was probably induced by hornblende dehydration, with melts having segregated into the irregular, diffuse, coarse-grained tonalitic regions. Mashapaug, Connecticut, USA.

Figure 21.7 Granodiorite gneiss containing small, discontinuous pegmatitic leucosomes that have separated into extensional fractures. The leucosomes have been elongated and boudinaged as a result of later deformation, making them approximately parallel to the weak gneissic layering. The red arrows point to a deformed dike that is cut by the leucosomes, and so pre-dates them. A close-up of one of the leucosomes is shown in Figure 21.8. Midøy, Midsund, Møre og Romsdal, Norway.

Figure 21.8 Close-up of one of the pegmatitic leucosomes in Figure 21.7. The assemblages in both the segregations and the enclosing gneiss are the same: microcline – quartz – plagioclase – biotite. The segregations, however, have K-feldspar and quartz in proportions that are higher, and biotite proportions lower, than in the host gneiss, approximating a granitic minimum-temperature melt. The field width is about 12 cm. Midøy, Midsund, Møre og Romsdal, Norway.

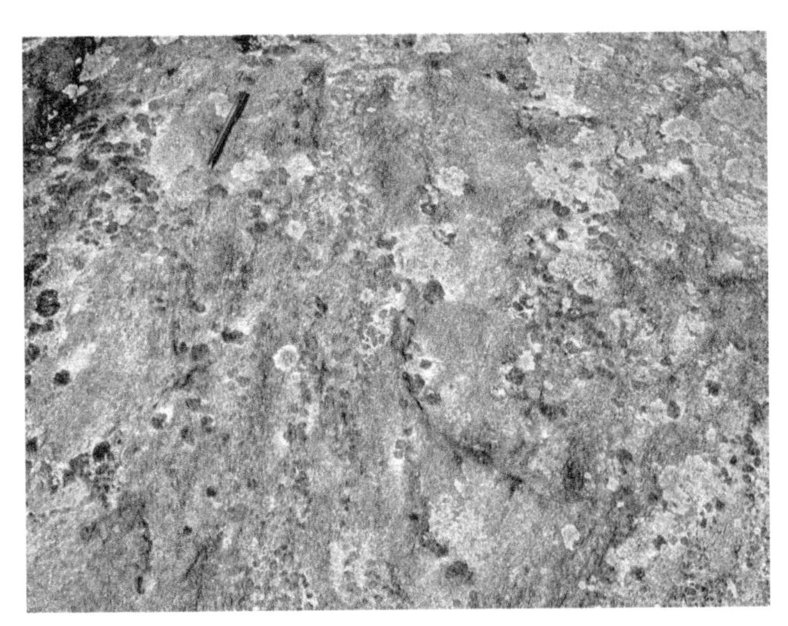

Figure 21.9 Coarse, irregular leucosomes developed in pelitic schist that was metamorphosed to granulite facies conditions. The fine-grained host rock contains the assemblage quartz – K-feldspar – biotite – garnet – cordierite – sillimanite. Garnets in the gray matrix rock are relatively small, 1–3 mm across, and sillimanite is abundant. The coarse leucosomes have the same assemblage, but with more quartz and K-feldspar, and much less sillimanite and biotite. The leucosomes also have particularly large garnets, up to 2.5 cm across. It appears that the presence of silicate liquid allowed the garnets to grow larger than those in the surrounding matrix. Gilbertville, Massachusetts, USA.

Figure 21.10 Kyanite-bearing pegmatite layer in a partially melted muscovite – biotite – garnet – kyanite schist. The silicate liquid segregated into extensional fractures, but later deformation extended and boudinaged the leucosomes, making the resulting fragments essentially parallel to the mica foliation. The pegmatites contain K-feldspar and abundant quartz, whereas the schist itself has less quartz and little K-feldspar. Lepsøy, Nordøyane, Møre og Romsdal, Norway.

Figure 21.11 Gray tonalitic gneiss hosting pre-metamorphic xenoliths of black amphibolite. Cutting the tonalite are light-gray leucosomes that were partial melts derived from this and nearby tonalitic gneisses. There are at least two episodes of leucosome emplacement, as indicated by crosscutting relationships (inside red ellipses) and folded leucosomes (yellow arrows) compared to adjacent, relatively undeformed leucosomes (white arrow). Notice that some of the leucosomes contain small fragments of the gneiss (blue arrows), and many have thin, black, biotite-rich selvages (black arrows). These are interpreted to indicate successive emplacement of small melt volumes into progressively opening fractures, with biotite preferentially crystallizing on the fracture walls. Quabbin Reservoir, Massachusetts, USA.

Figure 21.12 Light-colored tonalitic partial melts (leucosomes) that cut darker gray tonalitic gneiss. Several generations of melt injection have taken place, as indicated by different dike orientations, different grain sizes, and the numerous, dark, biotite-rich selvages within the leucosomes. Quabbin Reservoir, Massachusetts, USA.

Figure 21.13 Coarse, dark-gray to black boudins of hornblendite, enclosed in gray tonalitic gneiss. Parts of the mafic rock have been disrupted by invading white tonalitic leucosomes that were mobilized as partial melt from the surrounding gneiss. The presence of grain boundary silicate liquid at lithostatic pressure probably weakened the hornblendite, which fragmented during ongoing deformation. Fragmentation allowed fine-scale infiltration of the liquid into the disrupted boudins. Quabbin Reservoir, Massachusetts, USA.

Figure 21.14 Coarse, white pegmatite dikes (P) cutting light-gray tonalitic gneiss and dark-gray amphibolite. The K-feldspar-bearing pegmatites cannot have come from the gneisses or amphibolites, and so were probably derived from partial melting of nearby muscovite – K-feldspar – sillimanite schist (Tracy, 1978). Melting was induced by muscovite dehydration, and the liquid traveled perhaps tens of meters along fractures out of the schists into these rocks. The tonalitic gneisses also have thin, approximately layer-parallel leucosomes that were apparently derived from the felsic gneisses themselves (red arrows). The bright white material is snow. Quabbin Reservoir, Massachusetts, USA.

Figure 21.15 Thin and thick pink pegmatitic leucosomes in granitic gneiss. The thinner leucosomes have smooth surfaces and appear to have been highly deformed to be parallel to the gneissic layering (one example indicated with a yellow arrow). The thicker leucosomes are younger and, though deformed, crosscut the gneissic layering at a low angle. Some of the thick leucosomes have melanosomes on their margins, but they are discontinuous and vary widely in thickness, from thick (red arrow) to nonexistent (blue arrows). It is possible that the selvages are unrelated to pegmatite origin. Instead, the pegmatites may merely have intruded along a fracture that formed in weak, biotite-rich layers in the gneiss. Hemnskjela, Sør Trøndelag, Norway.

Figure 21.16 Leucosomes that vary in the sharpness of their contacts with the host tonalitic gneiss. Some contacts have marginal melanosomes, some of which are indicated by red arrows. Those may represent different amounts of melt-restite separation from the originally melted rock immediately adjacent to the leucosomes. Most of the leucosomes are parallel to the gneissic layering, indicating that melting, melt segregation, and melt crystallization occurred before development of the folds that are evident here. Kopparen, Fosen Peninsula, Sør Trøndelag, Norway.

Figure 21.17 Migmatitic dioritic gneiss with at least two episodes of leucosome development. The earliest leucosomes are visible as thin white to pink layers in the dark-gray gneiss. The gneiss and leucosomes were folded, producing approximately parallel light and dark layers. The second, younger set of leucosomes are also white to pink, but occur as crosscutting dikes and irregular, anastomosing patches. The layer-parallel leucosomes may have been derived from the gneiss, but the crosscutting leucosomes seem to have been introduced from the outside. Lepsøy, Nordøyane, Møre og Romsdal, Norway.

Chapter 22

Retrograde metamorphism

Despite what was said in Chapter 1 about the peak metamorphic temperature assemblages usually being preserved, retrograde metamorphic effects are actually almost ubiquitous in medium- to high-grade regionally metamorphosed rocks. This is because it takes a long time for those rocks to reach the surface, so there is a lot of time for them to partially reequilibrate to the new P-T conditions, and to be exposed to fluids, as the rocks are cooling and decompressing (Fig. 22.1A).

The most commonly observed retrograde metamorphic effects involve partial replacement of one or more prograde minerals by others, to produce new, retrograde assemblages that are characteristic of lower metamorphic grades. Common examples are replacement of garnet by chlorite, biotite by chlorite and a K-bearing mineral, aluminosilicates and K-feldspar by muscovite, and cordierite and biotite by chlorite and muscovite. Many retrograde reactions are driven by the availability of H_2O, which permits the reverse of equivalent prograde reactions to take place. For example, the prograde dehydration reaction,

$$\underset{\text{muscovite}}{KAl_2Si_3AlO_{10}(OH)_2} + \underset{\text{quartz}}{SiO_2} = \underset{\text{sillimanite}}{Al_2SiO_5} + \underset{\text{K-feldspar}}{KAlSi_3O_8} + \underset{\text{fluid,}}{H_2O} \tag{22.1}$$

can be reversed in the presence of aqueous fluid, under the right P-T conditions (Fig. 22.1B, C), to become a retrograde hydration reaction. The aqueous fluid source can be external to the retrograded rock, perhaps derived from nearby plutons, or fluids migrating upward from decompressing, dehydrating rocks below (Fig. 22.1D, E). The H_2O may also be intrinsic to the rock, as grain boundary metamorphic fluid that did not escape (Fig. 22.1F), or fluids that un-mix from solution in nominally anhydrous minerals such as quartz (Fig. 22.1G; Spear and Selverstone, 1983). It is easiest to determine that retrograde reactions took place if remnants of the prograde assemblage remain: cores of garnets with rims of chlorite (Fig. 22.1H), sillimanite inclusions in muscovite and quartz (22.1C), or islands of omphacite inside diopside – plagioclase symplectites (Figs. 11.2C, D). In the absence of incomplete retrograde reaction textures, other clues need to be looked for. Such clues may include spherical or even faceted masses of polycrystalline chlorite that indicate pseudomorphic replacement of garnet, or epidote-albite intergrowths that may indicate replacement of plagioclase.

At the very least, retrograde effects usually involve exchange of chemical components between adjacent unlike minerals (Fig. 22I, J). At low temperatures, the mineral parts affected by component exchange may be only nanometers thick. Cooling from high temperatures, however, can allow exchange and solid state diffusive transport of some

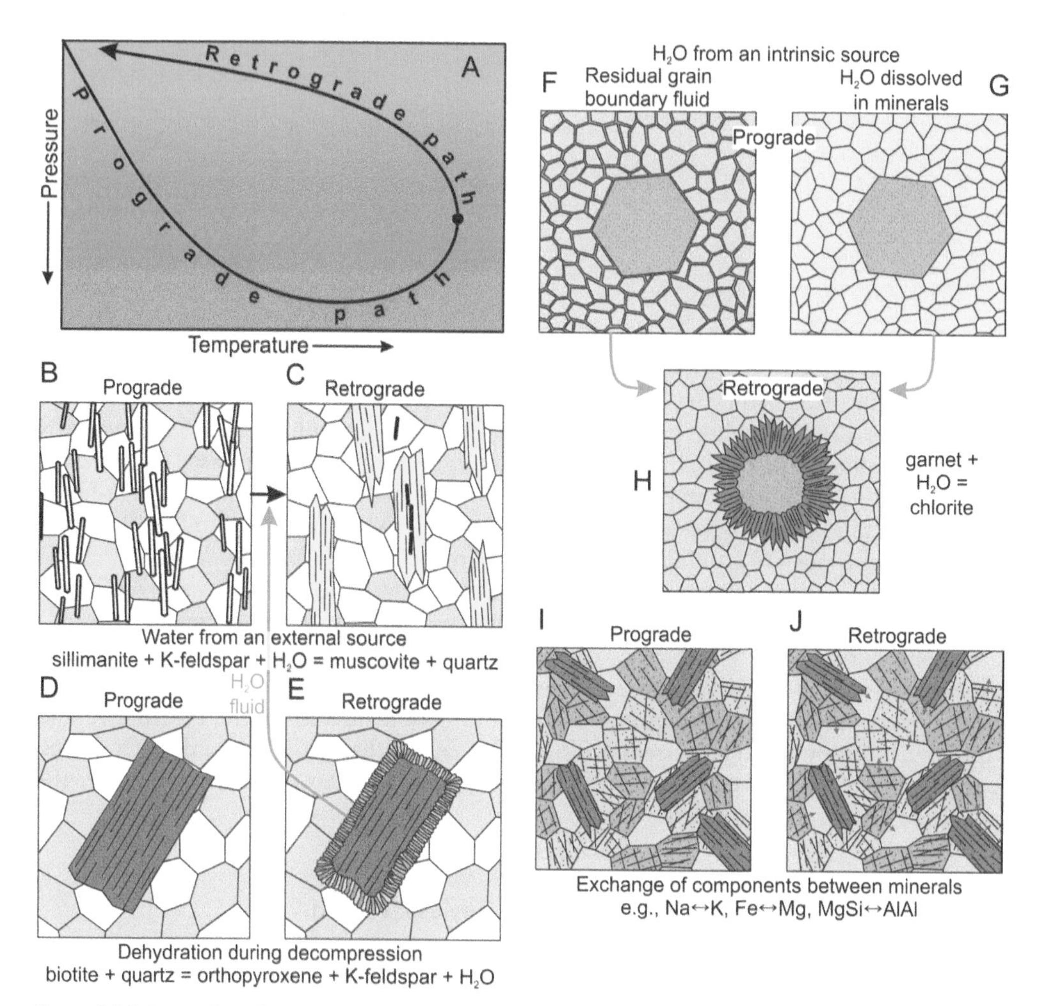

Figure 22.1 Examples of some processes that result in retrograde metamorphic effects. A) Schematic prograde and retrograde paths for a rock passing through P-T space. The black dot is the point of maximum temperature, which divides the prograde and retrograde paths. B, C) Under retrograde conditions, available aqueous fluid can allow hydration reactions to run. In this case, the reaction is the reverse of prograde reaction 22.1. Traces of prograde sillimanite remain protected inside muscovite and quartz, indicating the former prograde assemblage. D, E) One possible source of aqueous fluids for retrograde metamorphism is the dehydration of hydrous minerals deeper in the crust, as uplift and unroofing causes decompression. In this case, biotite dehydration releases aqueous fluid that migrates to higher levels, permitting the retrograde hydration reaction in B and C to run. In addition to external fluid sources, there may also be internal sources if fluid produced by prograde reactions did not completely escape. Some may reside as grain boundary fluid (F), or dissolved in nominally anhydrous minerals (G; Spear and Selverstone, 1983). In either case, the excess fluid is consumed under retrograde conditions where more hydrous mineral assemblages are stable (H). I, J) Less obvious than mineralogical change is the exchange of chemical components between different phases (red arrows). The results of this exchange depends in part on the distance chemical components can diffuse within the solid crystals and along grain boundaries. Commonly, the rate of diffusion in garnet, for example, is slow enough so that only the rims are affected. In biotite, by comparison, diffusion is fast enough so that crystals may be homogeneous from internal solid state diffusion. Such composition changes must be considered when collecting samples for P-T determinations and related work. Although examples D, E and I, J may take place as the rock follows the retrograde decompression and cooling path, they are not generally considered retrograde metamorphism in the sense of resetting the mineral assemblages to those of lower P-T conditions, as are B, C and F-H.

chemical components over millimeters or centimeters into adjacent minerals. Exchange of components in this way is generally invisible at the outcrop and in thin section, and requires microbeam analytical methods to resolve. Such composition changes must be understood before meaningful estimates of metamorphic pressures and temperatures can be made.

There are several other processes that occur along the retrograde metamorphic path (Figs. 1.3, 22.1A) that are not generally considered to be retrograde metamorphism as such. One of these is the unmixing of solid solutions into two or more stable, low-temperature phases. Unmixing of K-Na feldspar solid solutions into distinct intergrowths of K-rich and Na-rich feldspars (perthite) is perhaps the best known example (barely visible in Figs. 21.2A, B, more visible in Fig. 8.2F). Other examples include unmixing of hornblende into hornblende and cummingtonite lamellae, orthoamphibole into gedrite and anthophyllite lamellae, and ortho- and clinopyroxenes into one another. While generally on too small a scale to be visible on the outcrop, if the exsolution lamellae are evenly spaced and close to a light wavelength apart, internal colors from light diffracted from the lamellae may be seen. Another retrograde path feature is some blue quartz, which is caused by light scattering from sub-micron scale titanium oxide crystals that have unmixed from high temperature solid solution.

One might wonder what the difference is between retrograde metamorphism and fluid metasomatism (Chapter 20). The answer is not entirely straightforward because of variations in how the terms are used in practice by different people, and because the terminology represents an imperfect way to classify a range of natural processes. Perhaps simplest, typical retrograde metamorphism involves transforming high grade metamorphic assemblages into assemblages characteristic of lower metamorphic grades. This transformation may involve an influx of fluids (Figs. 22.1B, C, and 22.2A-F), or it may not (e.g., eclogites, Figs. 11.2A-D, ignoring minor hornblende, and Figs. 22.1D, E and 22.2G, H). During retrograde metamorphism there is generally no pervasive change in chemical composition, other than fluid components like H_2O and CO_2. Retrograde metamorphism also generally refers to broad areas rather than to local effects, around small fractures, for example. Metasomatism, in contrast, is usually thought of as involving limited areas, with important changes to the rock chemical composition besides just the fluid components. Commonly, metasomatism transforms one rock into a completely different rock (e.g., Figs. 20.1, 21.5, 21.6, 21.10, 21.12), rather than simply changing the mineral assemblage from that characteristic of one grade to another. As was mentioned before, it is best to describe what you actually observe, rather than relying on the definition of a word that may change in your mind over time. That way, your field notes remind you of what you actually saw.

Like so much of metamorphic geology, identifying retrograde metamorphic effects requires drawing on a wide range of knowledge and experience. The more you know about mineral formulae, mineral identification, and possible reactions, the more easily metamorphic textures can be interpreted in plausible ways in the field.

The thin section images first show two classic examples of chlorite grade replacement of higher-grade prograde assemblages in pelitic schists (Fig. 22.2A-D). The next example shows retrograde replacement of orthopyroxene by cummingtonite in a migmatite leucosome (Fig. 22.2E, F). Figures 22.2G and H shows a retrograde-path reaction texture that did not involve fluids, but might not be called retrograde metamorphism. The field photos also start with chlorite grade retrograde assemblages replacing higher-grade, prograde assemblages (Figs. 22.3–22.6). Those are followed by two examples of retrograde

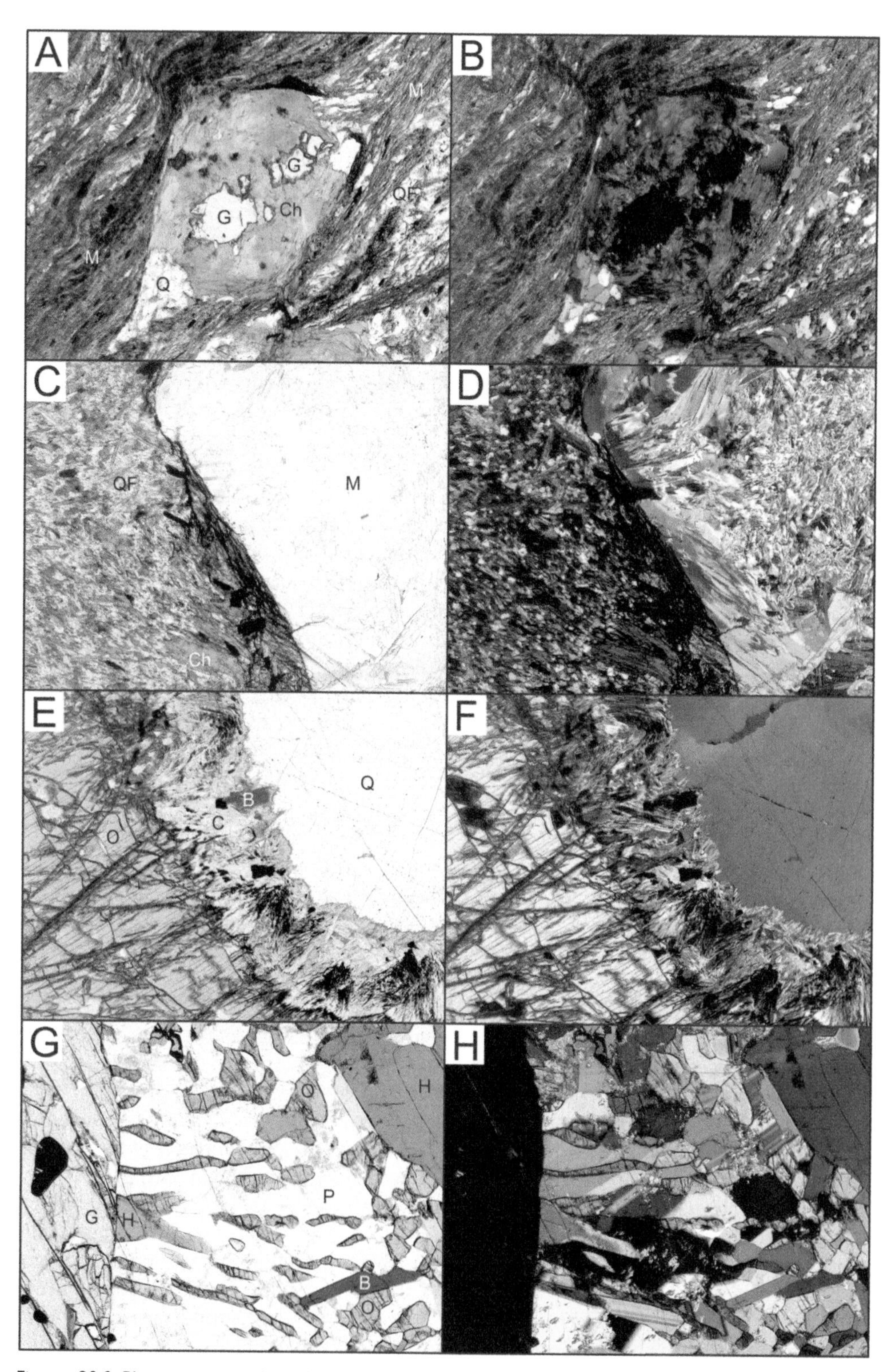

Figure 22.2 Photomicrographs of thin sections showing evidence of retrograde metamorphic reactions. All the images on the left are in plane-polarized light, and all on the right are the

Figure 22.2 (Continued) same fields in cross-polarized light. All fields are 4 mm across. A, B) Garnet in a fine-grained pelitic schist, partially replaced by retrograde chlorite (garnet + H_2O = chlorite + quartz). New Salem, Massachusetts, USA. C, D) A large andalusite porphyroblast in pelitic schist that has been entirely replaced by muscovite (andalusite + biotite + quartz + H_2O = muscovite + chlorite). Chlorite in the matrix is retrograde by this same reaction, pseudomorphing biotite. East Rochester Center, New Hampshire, USA. E, F) Orthopyroxene in a tonalitic partial melt segregation, partially replaced by a rim of fine-grained cummingtonite (orthopyroxene + quartz + H_2O = cummingtonite). Ware, Massachusetts, USA. G, H) Retrograde path reaction between garnet and the surrounding matrix to produce a symplectite, approximately by the reaction: garnet + plagioclase = orthopyroxene + more calcic plagioclase. This reaction took place during decompression at high temperature. Some might object to calling the reaction retrograde because the reaction doesn't produce hydrous minerals, or an assemblage characteristic of lower temperatures. However, it can be thought of as being like the retrograde symplectites seen in eclogites, which form in much the same way (Fig. 11.2, if one ignores the hornblende). North River, Adirondacks, New York, USA. Reactions are approximate and not balanced. Abbreviations: B, biotite; C, cummingtonite; Ch, chlorite; G, garnet; H, hornblende; M, muscovite; O, orthopyroxene; P, plagioclase; Q, quartz; QF, quartz and feldspar.

hydration of almost anhydrous granulite facies rocks (Fig. 22.7, 22.8). Figure 22.9 shows retrograde replacement of a crystal in a partial melt segregation, analogous to retrograded minerals in Figure 21.1Ae, and the same as Figure 22.2E, F. Figure 22.10 shows a retrograde reaction texture and assemblage that did not involve significant amounts of fluid, like Figures 22.2G, H. Lastly, Figure 22.11, shows an example of exsolution that can be seen in the field (retrograde path process, but not retrograde metamorphism).

Figure 22.3 This rock is a garnet grade schist containing garnet, muscovite, biotite, and chlorite. Elongate green chlorite crystals are visible in the matrix. The prograde garnets are now partially replaced by chlorite rims (red arrows point to where they are thickest). The matrix chlorite is Mg-rich, but the retrograde chlorite on garnet is more Fe-rich, having grown principally from components derived from the Fe-rich garnet. The field width is about 12 mm. Figures 22.2A and B are a more extreme example of the same thing. Bridgewater Corners, Vermont, USA.

Figure 22.4 Pink andalusite crystals partially replaced by white rims of coarse-grained retrograde muscovite. The dark crystals above and below the large, center andalusite grain (red arrows) are not biotite, but muscovite, seen edge-on. In that orientation, muscovite is more transparent, allowing the shadowed, dark, deep crystal interiors to be seen. This rock also has fine-grained, fibrous sillimanite (yellow arrows). The rock apparently passed from the andalusite stability field into the sillimanite field during prograde metamorphism, growing some sillimanite but leaving considerable metastable andalusite. Later, the rock passed back into the andalusite field, and aqueous fluid permitted retrograde muscovite to partially replace both andalusite and sillimanite. Andover, Maine, USA.

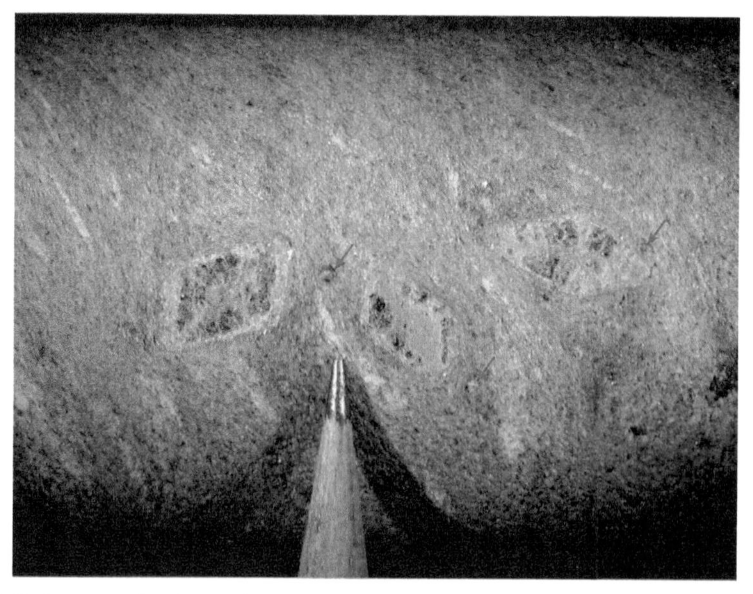

Figure 22.5 Drill core of pelitic schist with orange-brown staurolite porphyroblasts. The porphyroblasts have been partially replaced by a gray mixture of chlorite and muscovite. The replacement is irregular, affecting both the staurolite rims and cores, but leaving the intermediate parts. Based on examination of intact prograde staurolite in nearby rocks, it turns out that staurolite cores were rich in quartz inclusions. Not only did the inclusions increase the grain boundary surface area along which fluid could migrate, but the extra quartz allowed faster progress of the quartz-consuming retrograde reaction (approximately staurolite + biotite + quartz +

Figure 22.5 (Continued) H_2O = muscovite + chlorite). The red garnets (red arrows indicate some) are fresh, without retrograde chlorite rims. New Salem, Massachusetts, USA.

Figure 22.6 Pelitic schist containing retrograde muscovite pseudomorphs after andalusite (dark gray, red arrows) and muscovite – chlorite pseudomorphs after staurolite (brownish-gray, yellow arrows). The pseudomorphs after staurolite, in particular, retain their original euhedral shapes. The pseudomorph after andalusite in the upper-right corner is lighter-colored in its center. It is only the rims that are dark-gray, because that is where graphite was pushed to the side as the porphyroblasts grew. Figures 22.2C and D shows a muscovite pseudomorph after andalusite that is very much like those shown here. Sterling, Massachusetts, USA.

Figure 22.7 This rock was originally a two-pyroxene – garnet granulite. Along shear zones (at the horizon of the marker pen and near the image bottom), aqueous fluid had access to the rock and turned the granulite facies assemblage to amphibolite, and allowed some black hornblende to grow in the immediately adjacent granulite. Because fluid access was largely limited to shear zones, is this an example of metasomatism instead of retrograde metamorphism? Maybe, though replacement of a granulite facies assemblage by an amphibolite facies assemblage is a retrograde process regardless of its scale. Garnets in the granulite are surrounded by thin, white, retrograde plagioclase-rich

Figure 22.7 (Continued) rims, visible in the inset. These rims resulted from a reaction between the garnet and surrounding anhydrous minerals during decompression, something like that shown in Figures 22.2G and H. Midøy, Midsund, Møre og Romsdal, Norway.

Figure 22.8 Coarse anorthosite that was strongly deformed and originally metamorphosed to granulite facies conditions. The recrystallized plagioclase grains are now less than 1 cm across, much smaller than they probably were originally, judging from the adjacent pyroxenes. The orthopyroxene (brown) and clinopyroxene (black) crystals are up to 15 cm long, but have fine-grained, dark-green to black, retrograde rims of hornblende and anthophyllite. Kaupanger, Sogn og Fjordane, Norway.

Figure 22.9 Orthopyroxene in a crystallized tonalitic partial melt segregation, in amphibolite that was metamorphosed to upper amphibolite facies conditions. Here, the orthopyroxene (rusty-weathering cores in the two large grains) has been partially replaced by thick black rims of fine-grained cummingtonite, the product of the retrograde hydration reaction: orthopyroxene + quartz + H_2O = cummingtonite. The much smaller orthopyroxene crystals in the host amphibolite have been entirely replaced by fine-grained cummingtonite. These crystals are from the same rock as Figure 22.2E and F. Ware, Massachusetts, USA.

Figure 22.10 Orthopyroxene-bearing garnet amphibolite, in which the garnets are surrounded by 2–4 mm thick plagioclase – orthopyroxene symplectite rims, like those seen in Figure 22.2G and H. The reaction is somewhat complicated but it can be approximated by: garnet + plagioclase = orthopyroxene + more calcic plagioclase. The symplectite formed on the retrograde path, during decompression, soon after peak prograde metamorphic temperatures. North Creek, Adirondacks, New York, USA.

Figure 22.11 An example of exsolution as an intra-crystalline retrograde path process. This rock is a cordierite – gedrite gneiss, metamorphosed to kyanite grade. The white patches are cordierite, the brown lamellar mineral, particularly visible in the lower left, is gedrite, and the red specks are rutile. The blue areas to the upper right are a play of colors from inside the gedrite crystals. The colors result from exsolution of parallel, alternating anthophyllite and gedrite lamellae, during cooling from the originally homogeneous solid solution. The exsolution lamellae are regularly spaced, about a light wavelength apart, and diffract blue light. This effect is visible in the field by eye and with a hand lens, and is best seen on wet, fresh surfaces such as this one. The image field width is about 8 mm. What was the protolith for a gedrite – cordierite rock? It was probably a basalt that had been hydrothermally altered at high temperature with circulating sea water. The important composition change involved the exchange of calcium in the rock for magnesium in sea water, resulting in an Mg-rich, Ca-poor, peraluminous protolith (e.g., Schumacher, 1988). Richmond, New Hampshire, USA.

Part 3: Relict pre-metamorphic features

Chapter 23

Relict sedimentary features

Sediments and sedimentary rocks host an enormous range of features unique to the materials they are made of, the processes that deposit them, and changes that occur after deposition but before lithification. These features can include soft sediment deformation (Fig. 23.1A), crossbeds (Fig. 23.1B), mineralogical or size-graded beds (Fig. 23.1C), and intra-unit (Fig. 23.1D) or larger-scale angular unconformities. Other features may include bed forms like ripple marks and the shape and mineralogy of detrital mineral grains.

Preservation of such features through metamorphism depends on many things, including the original lithologic or mineralogical contrasts between the protolith sedimentary features, the metamorphic grade, and the amount and style of deformation. Generally the less the lithologic contrast, the higher the metamorphic grade, and the more severe the deformation, the less likely original sedimentary features will be identifiable. No surprise there. Finding original sedimentary features can be immensely important for use as stratigraphic top indicators, evidence for the depositional environment, and evidence for the original sediment source regions. All of these make it worthwhile to look for relict sedimentary features during field work.

For example, in complexly folded metamorphic terranes, relict sedimentary top indicators can help unravel the structural complexities (e.g., Thompson, 1988). Age dates of individual detrital zircon grains from quartzites can help determine the original sand source terrane (was the source Laurentia or Gondwana, Baltica or Amazonia?; e.g., Wintsch et al., 2007). Even surviving bed forms can help determine if a unit was deposited in shallow water (e.g., river valley or coastline) or deep water (e.g., continental rise). While the identification of sedimentary structures in metamorphic rocks can be difficult, even controversial, try not to admit that you are wrong too soon.

Thin sections are generally on too small a scale to show most relict sedimentary features, except detrital grains. Figure 23.2 shows detrital zircon and feldspar in two low grade quartzites. The field photos first show soft sediment deformation features in folded, chlorite grade rocks (Figs. 23.3, 23.4), followed by crossbedded quartzites at grades ranging from greenschist to epidote amphibolite facies (Figs. 23.5–23.7). Following that is a probable example of a scour and fill structure that survived sillimanite grade metamorphic conditions (Fig. 23.8). Figures 23.9 and 23.10 show graded beds, metamorphosed to amphibolite facies conditions, and interpreted to be deep marine turbidity current deposits. Figure 23.11 shows some well-preserved flute casts on the underside of a turbidity current deposit in slate. Figure 23.12 shows a remarkable, and remarkably small, contact metamorphosed unconformity. Lastly, Figure 23.13 shows the metamorphic product of shale deposition under highly anoxic conditions.

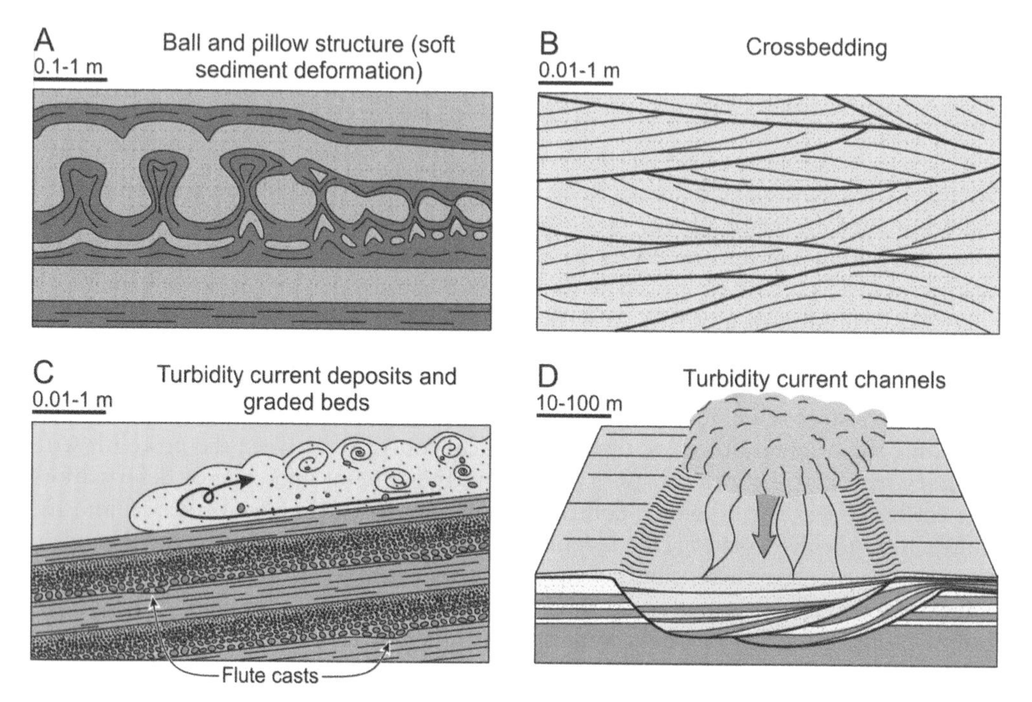

Figure 23.1 Examples of some sedimentary structures that might be seen in metamorphic rocks. A) Soft sediment deformation, in this case ball-and-pillow structures which form as a result of density instability. Dense sand is shown interbedded with less dense clay or silt. Some disturbance triggers the sand to sink into the fine-grained layer, and conversely the fine-grained material to rise into the sand. The sequence from left to right shows increasing disruption of the layers. Such structures can involve other kinds of sediments as well, such as carbonate sand and volcaniclastics. B) Crossbeds typical of fluvial or near-shore sandstone, but also found in other sedimentary environments. C) Turbidity current deposits, which commonly include size-graded sand beds interlayered with shale. Flute casts and other sole marks may occur on sand bed bottoms. D) Turbidity current channel scour-and-fill deposits, somewhat like the smaller scale crossbeds in B. Some structures that develop during metamorphism can resemble sedimentary structures (e.g., Fig. 3.10), so it is important to carefully consider metamorphic alternatives so as not to reach incorrect conclusions. Scales show typical size ranges for the features. A is adapted from Ghosh et al. (2012), B from Grabau (1920, p. 454), C from Bentley, 2009, and D from Lien et al. (2003).

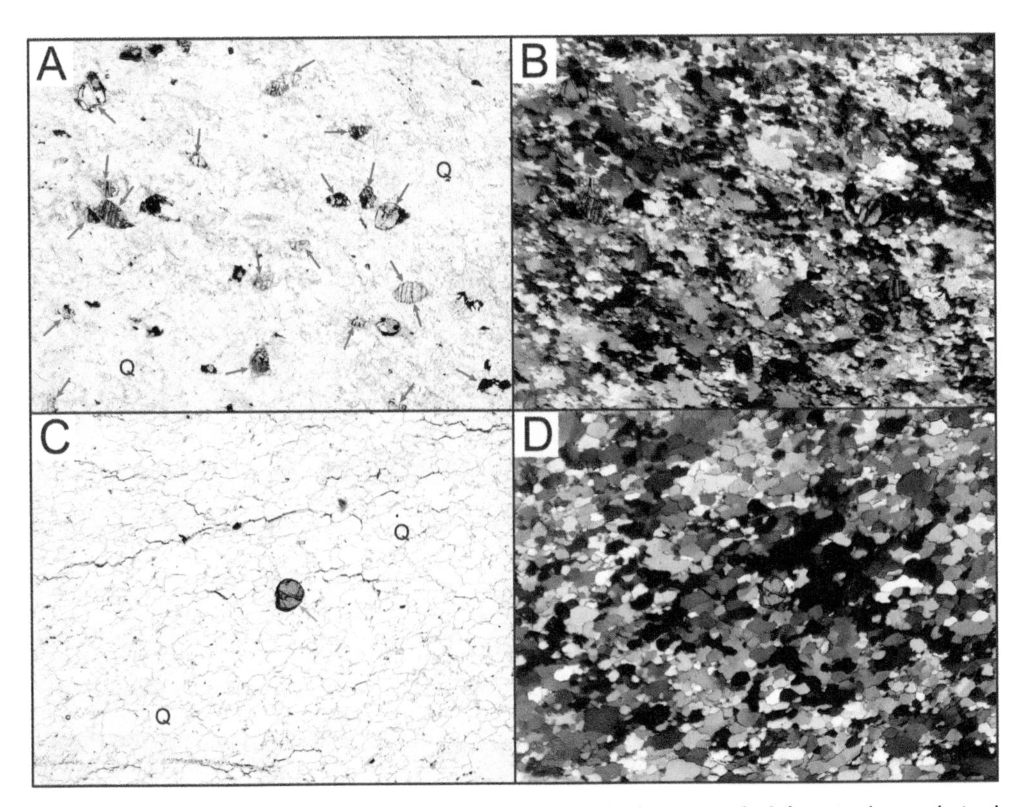

Figure 23.2 Photomicrographs of relict detrital grains in quartzite. Images on the left are in plane-polarized light and those on the right are the same fields in cross-polarized light. All field widths are 4 mm. A, B) Feldspar grains (red arrows) can survive to moderate metamorphic grades (here, biotite grade). Feldspars are more deformation-resistant than quartz under mid-crustal conditions, so they tend to maintain their original grain size and rounded detrital shapes better than quartz. However, if they react with aqueous fluid to form white micas, they can easily lose their detrital identity. Bennington, Vermont, USA. C, D) Rounded detrital zircon grain (blue arrow) in quartzite, here at garnet grade. Of the ZTR mineral group (zircon, tourmaline, rutile, minerals that are exceptionally weathering-resistant in addition to quartz), zircon is generally the most common. If zircon, tourmaline, or rutile are absent in a quartzite, then the protolith sediment source area probably did not have extensive exposures of deeply eroded igneous or metamorphic rocks. Williamstown, Massachusetts, USA. Can these features be seen with a hand lens? Yes, in some cases, but it requires careful observation and at least a few thin sections to help train the eye (e.g., feldspar grains in quartzite, Fig. 3.4).

Figure 23.3 Sedimentary layers disrupted by soft sediment deformation, here metamorphosed to greenschist facies conditions. The contorted light-colored layer immediately above the pencil is metamorphosed siltstone, which contains blocks of darker metamorphosed sandstone broken from the layers above and below. The adjacent layers seem not to be disturbed, suggesting that deformation was related to bedding-parallel sliding. Blekkpynten, Fættenfjord, Nord Trøndelag, Norway.

Figure 23.4 Contorted layers from the same location as Figure 23.3. Here you can see the disrupted top of a dark-colored, relatively coarse-grained metamorphosed sandstone layer (below the pencil), and overlying, folded finer-grained rock (behind and above the pencil). Asymmetric folds indicate a top-left shear sense. The fact that only this one bed is deformed, and the folds are in a completely different orientation from surrounding large-scale folds and foliation, supports the idea of soft sediment deformation. Fractures below and to the left of the pencil are filled with calcite, much of which has dissolved away. Blekkpynten, Fættenfjord, Nord Trøndelag, Norway.

Figure 23.5 Crossbedding in sandstone layers, metamorphosed to greenschist facies conditions and cut by quartz veins. The rocks are not highly deformed, and are at low grade, so the identification of crossbedding is relatively easy. Red lines show the boundaries of crossbed sets, and blue lines are parallel to the crossbedding itself. Caucomgomoc Lake, Maine, USA.

Figure 23.6 Crossbeds in clean quartzite, metamorphosed to biotite grade conditions. White, dotted lines highlight some of the crossbeds, most of which are tabular sets of foreset beds. The bed sets dip slightly to the left, approximately parallel to the knife, whereas the crossbeds themselves dip both left and right. The yellow arrows indicate low-angle bottomset beds, and the red T indicates an apparent trough. The same rock can be seen in thin section in Figures 22.2A and B. Bennington, Vermont, USA.

Figure 23.7 Crossbedded sandstone metamorphosed to epidote amphibolite facies conditions. Red lines show the boundaries between crossbed sets, and blue lines are approximately parallel to the crossbeds themselves. This image is from a horizontal surface, so the beds are nearly vertical. Alsen, Jämtland, Sweden.

Figure 23.8 Possible scour-and-fill structure in quartzite and muscovite schist, metamorphosed to sillimanite – muscovite grade. Dotted lines highlight the contacts. Quartzite and schist layers on the lower-right are truncated against left-tilted layers in the center and left. The tilted layers are thicker to the left, and so become progressively more parallel to the layers on the right, upwards on the outcrop (also stratigraphically upward as indicated by graded quartzite layers outside the field of this image). On field trips to this stop, roughly half of the geologists have thought that this was indeed a metamorphosed sedimentary structure, but the rest thought it was a metamorphic feature, maybe an annealed ductile fault along the truncation surface. To those who thought this was a sedimentary feature, it was interpreted to have been related to deep marine turbidity current channel processes (e.g., Fig. 23.1D). On the left, the layers are offset along a late brittle fault. Jaffrey, New Hampshire, USA.

Figure 23.9 Interlayered quartzite (red arrow) and schist (yellow arrow), formerly turbidite sandstone and shale, respectively, at garnet grade. The stratigraphic up direction is toward the top of this image, as indicated by the sharp contacts at the bottoms of the lighter-gray quartzite (white arrow), and diffuse tops where they grade into somewhat darker-gray, and now coarser-grained schist. Hoston, Sør Trøndelag, Norway.

Figure 23.10 Graded quartzite layers between coarse-grained muscovite – kyanite – staurolite – garnet schist. Stratigraphic tops are to the upper left. The original sandstones were grain size-graded from coarse at the sharp bed bottom contacts (yellow arrows), to fine at the tops (toward the upper left), with shale interbeds. During metamorphism, the shale turned to schist and grew porphyroblasts of garnet, staurolite, and kyanite. The quartzite, however, has much smaller grains than the schist, possibly not much different than in the original sandstone. The metamorphic rock grain sizes (coarse schist, shale protolith, fine quartzite, sandstone protolith) have therefore reversed compared to the original sediments. Huntington, Massachusetts, USA.

Figure 23.11 Slate quarry wall, showing rock faces broken along the slate cleavage. The cleavage face between the two red lines, remarkably, broke parallel to the original bedding. On that face there are numerous flute casts (yellow arrows point out two) that are exposed on the underside of a metamorphosed sandstone bed (invisible, behind the flute casts). The slate cleavage apparently wraps around the casts. These rocks are somewhat overturned here, so the stratigraphic top direction is somewhat downward into the quarry face. The current direction indicated by the flute casts is shown by the blue arrow. Brownville, Maine, USA.

Figure 23.12 Part of the contact between a gabbro pluton and its host rocks. The black rock in the lower part of the image is gabbro, the light-colored rock above is Archean gneiss, and the small lens of rock in between, just below center, was a Cretaceous sediment. Both the gneiss and sedimentary rock have been contact metamorphosed to magmatic temperatures. This image shows not only an igneous intrusive contact (white dotted line), but an unconformity (yellow dotted line), where the sedimentary rock protolith rested on Archean metamorphic rock. Sometimes good things come in small packages. Skaergaard intrusion, east Greenland.

Figure 23.13 Rusty-weathering, highly sulfidic pelitic schist, containing large porphyroclasts of pyrite (center). Sulfide weathering produces iron sulfates, ferric iron hydroxides, and sulfuric acid, all of which help cover up, decompose, or disaggregate the underlying rock. This rock is like that in Figure 2.11, but much more sulfidic. This schist was metamorphosed to granulite facies conditions, and has the assemblage quartz – K-feldspar – plagioclase – sillimanite – cordierite – biotite – pyrite – graphite – rutile – pyrrhotite. The fresh rock, which is very hard to get at through the thick, weathered, very smelly crust, is essentially white, or gray from graphite. Biotite ranges in color from light-orange to colorless, being Ti-bearing, but containing almost no Fe because it is bound in the sulfides rather than silicates. The high sulfide content of this rock indicates an anoxic sea floor depositional environment, a lot of organic material deposition, and highly active sulfur-reducing bacteria. West Brookfield, Massachusetts, USA.

Chapter 24

Fossils in metamorphic rocks

> *"No longer call New Hampshire Azoic. Silurian fossils discovered to-day."* C.H. Hitchcock, *1870 (telegram to the Dartmouth Scientific Association, New Hampshire, USA, quoted in Hitchcock, 1874)*

Fossils are among the most important parts of the geologic record. As an indication of their importance, the Phanerozoic geologic time scale is based almost entirely on the fossil record. Radiometric dates are wonderful, and certainly provide absolute ages for the various geologic periods and their subdivisions (e.g., Cohen et al., 2013), but not all rocks host radiometric systems that can be used to date rock formation. Some volcanic rocks have zircons or other minerals that are usable to date formation ages. Rocks like quartzites, in contrast, get their detrital zircons from their source areas. The youngest zircon in quartzite only gives a maximum age for the original sandstone deposit. Plutonic rocks like granites have radiometric systems that can give their formation age, but yield only a minimum age for the rocks they crosscut. Many rocks, like shales, limestones, and zircon-free volcanics, contain no usable radiometric systems that both date their depositional age and are likely to survive metamorphism. In the vast seas of radiometrically undatable metamorphic rocks that exist in the world, fossils are rare but valuable chronologic pins on the time scales of those regions.

As one might expect, most fossils are destroyed during metamorphism by deformation and recrystallization. In unusual circumstances, however, fossils can survive in recognizable form even to high metamorphic grade. At the very least, such circumstances generally require minimal deformation of the fossil-bearing rock. Fossil preservation might occur by happenstance, such as a situation in which the fossil-bearing rock is preserved in less-deformed fold hinges. Deformation might also be limited if reactions early in the metamorphic history produced minerals that strengthened the fossil-bearing rock, allowing it to resist deformation. For example, an impure dolomite might react early during metamorphism to form a calc-silicate assemblage strong enough to resist deformation, if surrounding rocks were more ductile (e.g., Boucot and Thompson, 1963; Boucot and Rumble, 1978; brachiopods in sillimanite grade calc-silicate rock). Another possibility is that the fossils themselves were originally stronger than the host rock (e.g., Elbert et al., 1988, apatite conodonts in garnet grade marble).

In the absence of dateable igneous material containing zircons, or other appropriate radiometric systems, fossils can provide valuable evidence for understanding the age, stratigraphy, and structure of the host metamorphic rocks, and their fossil assemblage similarities or differences with distant, or once-distant, continents. Geologists working in metamorphic terranes would do well to keep their eyes open for them. At the very least a new locality is probably worth a publication.

The field photos are arranged from low to high metamorphic grade. The material of the fossils themselves includes carbon plus imprints in slate (Fig. 24.1), traces of disturbed sedimentary material in phyllite (Figs. 24.2, 24.3), original carbonate (Figs. 21.4–24.6), and replacement of the original fossil material by calc-silicate minerals (Figs. 24.7–24.9).

Figure 24.1 This rock is transitional between sedimentary and metamorphic, and can be thought of as a low-grade slate. The rocks are folded, and parts of the outcrop have variable amounts of foliation and cleavage development. The outcrop also contains marble, which is thoroughly recrystallized from its limestone protolith. The fossil graptolites shown here are *Monograptus* (Ordovician), and are found in the least deformed parts of the rock. Änge, Jämtland, Sweden.

Figure 24.2 Meandering trace fossil tracks of *Helminthoidichnites* (Ordovician) on a cleavage and foliation surface that happens to be parallel to the nearly vertical bedding surface of this greenschist facies phyllite. The early cleavage is cut by a later, weak cleavage. Faint undulations on the early cleavage surface, parallel to the yellow line, are crenulations of the cleavage-cleavage intersection lineation. Blekkpynten, Fættenfjord, Nord Trøndelag, Norway.

Figure 24.3 Close-up of one of the trace fossil tracks on the cleavage surface in Figure 24.2, showing the paired grooves that outline the trace fossil track, indicated by the yellow arrows. The cleavage-cleavage intersection lineation, indicated by small crenulations, is parallel to the red line. Blekkpynten, Fættenfjord, Nord Trøndelag, Norway.

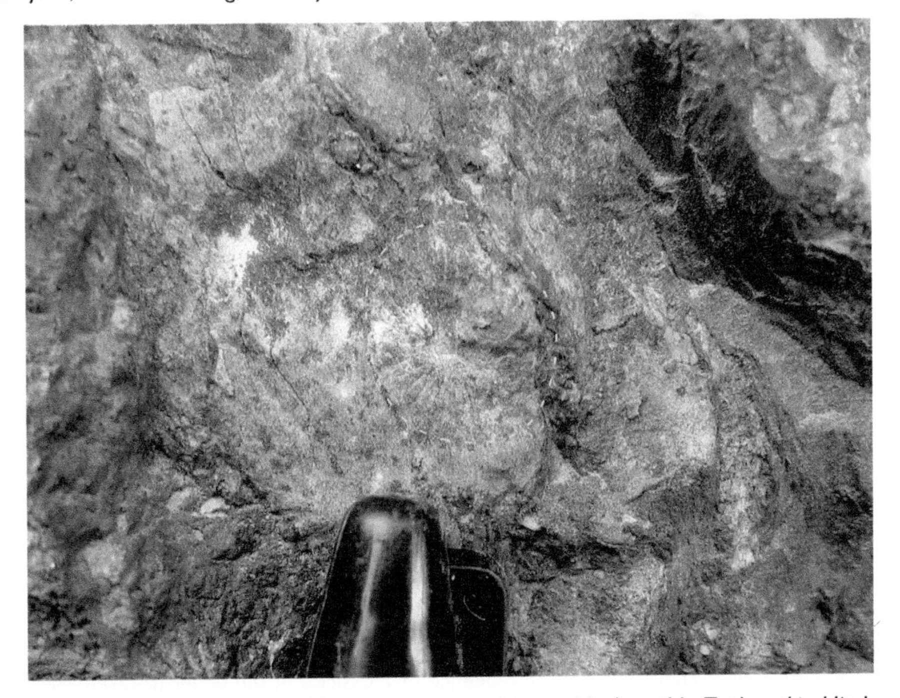

Figure 24.4 Transverse section of a Silurian rugose coral in marble (possibly *Zaphrenthis*, Hitchcock, 1874), metamorphosed to chlorite grade conditions. The radial septae are visible as light-gray lines within the yellow dashed line that outlines the fossil. Littleton, New Hampshire, USA.

Figure 24.5 Strongly deformed, lineated, impure marble, metamorphosed to greenschist facies conditions. The rock contains sparse, relatively undeformed crinoid ossicles. These crinoid fragments were probably preserved because they are each made of a single calcite crystal, which apparently resisted deformation better than the fine-grained, mica- and calcite-rich host rock. Ullån riverbed, Åre, Sweden.

Figure 24.6 Slightly deformed but otherwise intact sequence of 10 crinoid ossicles from the same locality as Figure 24.5. There is a smaller sequence of 7 ossicles in the upper left (red arrow). The yellow arrow points to what seems to be another ossicle set, but it is partly covered. The white arrow points to a rectangular object that may be a single ossicle in cross section. There are other suspiciously fossil-like objects scattered about as well. Ullån riverbed, Åre, Sweden.

Figure 24.7 Cross section of a fossil brachiopod (red arrow) in Silurian diopside – grossular – anorthite – quartz calc-silicate rock. The fossil itself is made up mostly of dark-green diopside rather than carbonate. The inset shows a *Spirifer*, with a plane slicing through it, approximating the upper valve cross section of the fossil seen in the larger image. Thin arcs and closed loops of diopside, pointed out with yellow arrows, may also be fossil forms. Montcalm, New Hampshire, USA.

Figure 24.8 This Devonian brachiopod shell bed is a calc-silicate layer at muscovite – sillimanite grade, as indicated by mineral assemblages in nearby schists. The fossils are made of variable proportions of calcite, quartz, and wollastonite, and are enclosed in the calc-silicate assemblage of quartz – grossular – diopside – hornblende – titanite – zoisite. The conversion of the protolith, possibly a sandy dolomitic limestone, to calc-silicates during early metamorphism apparently strengthened the shell bed, permitting it to survive later severe deformation. The calc-silicate layers are surrounded by quartzite, itself surrounded by pelitic schist, that were probably relatively ductile during metamorphism. Mt. Moosilauke, New Hampshire, USA.

Figure 24.9 Horizontal section through an upside-down stromatolite in Precambrian marble, metamorphosed to amphibolite facies conditions. The stromatolite was originally dolomitized and either included fine-grained quartz sand, or silica was introduced later, to react during metamorphism to diopside. The deformation-resistant diopside minimized deformation of this fossil, that occurs in severely deformed marble. Balmat, Adirondack Lowlands, New York, USA.

Chapter 25

Relict igneous features

Igneous rocks host a rich variety of features intrinsic to themselves, such as phenocrysts, cumulate layering, and pillow lavas. They also have features that show their relationships to the surrounding rocks, such as crosscutting contacts, chilled margins, and xenoliths. The rich variety of features, and impressive way in which they are formed, is indicated by the lavish illustrations found in many introductory geology textbooks (e.g., Marshak, 2012, p. 159).

With some imagination it is relatively easy to recognize many igneous features, despite the replacement of igneous with metamorphic mineralogy. However, the same cannot be said for the results of severe deformation, which can so distort the igneous features that they may no longer be recognizable. Dikes and their crosscutting relationships can be hard to distinguish from parallel layers of unknown origin as dike contacts approach parallelism with the features they crosscut (e.g., Fig. 8.1A, B). To cite another example, angular xenolith blocks can be deformed to appear in outcrop as parallel layers that may provide no hints to their origin (Fig. 8.1C, D). Unfortunately, one set of compositional layers of unknown origin can look much like another. Nonetheless, in some cases careful examination can lead to clues of how, and in what sort of environment, the original igneous rocks formed, and what relationship they have to their surroundings. Such interpretations of subtle, surviving igneous features ultimately help improve the overall geologic interpretation.

Figure 25.1 shows thin section photographs of a few metamorphosed igneous features that can be seen at that scale. Figures 25.1A and B show apparent igneous phenocrysts that have been pseudomorphed by metamorphic minerals, but retain the original euhedral phenocryst shapes. Figures 25.1C and D show a crystal interpreted to be a relict plagioclase phenocryst. Figures 25.1E and F show the effects of high-grade but static metamorphism of the chilled margin of a gabbro pluton.

The field photos start with examples of metamorphosed volcanic features including pillow lavas (Figs. 25.2–25.7, less-deformed to more-deformed), and felsic pyroclastics (Fig. 25.8, 25.9). Following that are crosscutting relationships in which dikes cut layered rocks (Figs. 25.10, 25.11) or relatively homogeneous metamorphosed plutonic rocks (Figs. 25.12, 25.13). Figure 25.14 is a metamorphosed composite dike, with textural, mineralogical, and chemical evidence for three magma batches having intruded along the same conduit. Figures 25.15 and 25.16 show igneous cumulate layering that has survived metamorphism, and Figures 25.17 and 25.18 show recognizable metamorphosed igneous phenocrysts. Figure 25.19 is of a preserved chilled margin at the edge of a gabbroic pluton, the same as that shown in Figures 25.1E and F. Lastly, Figures 25.20–25.22 show recognizable xenoliths in metamorphosed plutonic rocks.

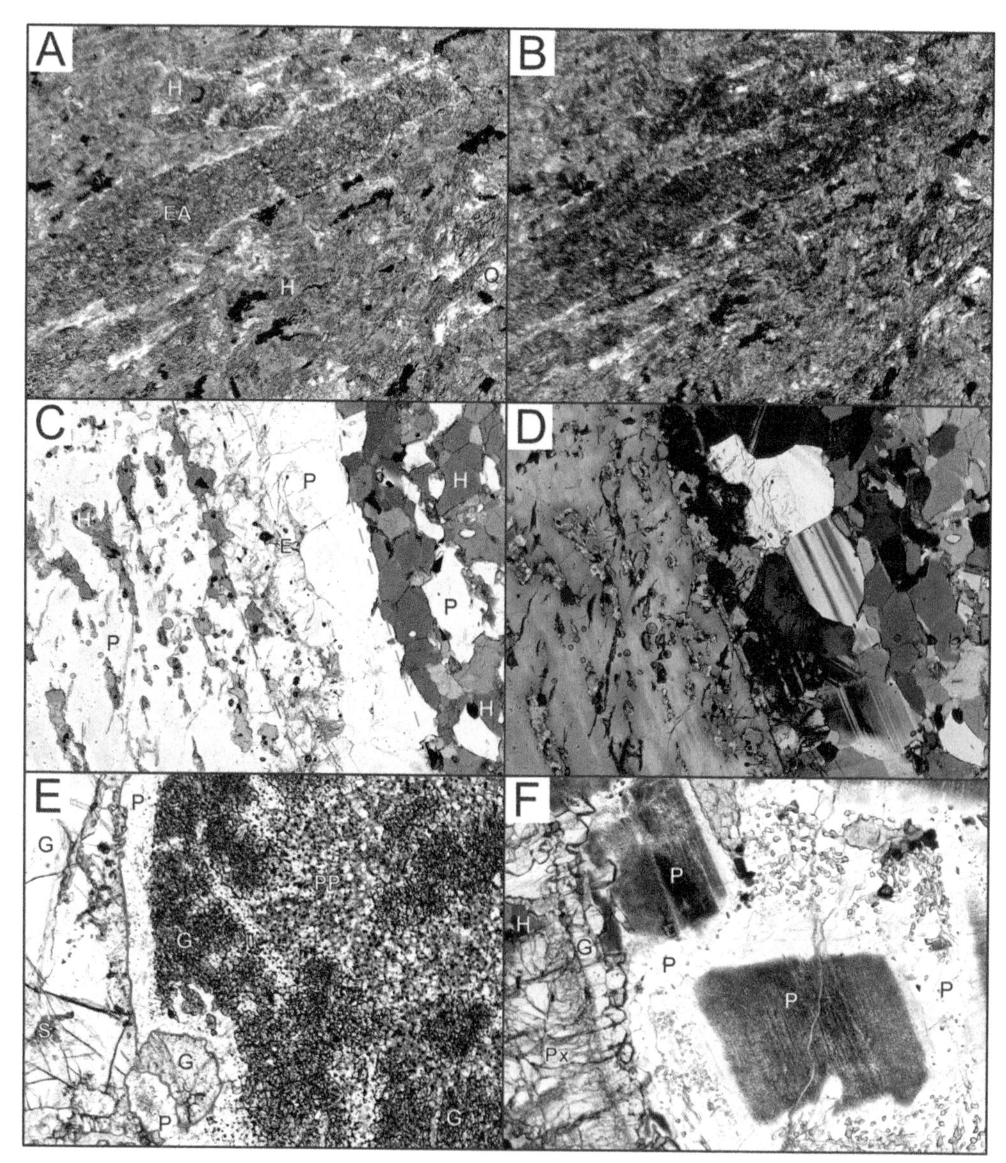

Figure 25.1 Thin section photomicrographs of small-scale igneous rock textures that have survived metamorphism. Images A and C are in plane-polarized light, and B and D are the same fields in cross-polarized light. Images E and F are different samples from the same rock body, both in plane-polarized light. All field widths are 4 mm. A, B) Sample from a boudin of a basaltic dike in quartzite, metamorphosed to epidote amphibolite facies conditions. The images show clinozoisite – albite pseudomorphs after plagioclase phenocrysts (cloudy brown in A, dark-gray in B). The phenocrysts are enclosed in a fine-grained, hornblende-rich matrix (greenish in A and B). In outcrop, the texture is similar to a porphyritic basalt (see Fig. 25.17). Dryna River, Oppland, Sør Trøndelag, Norway. C, D) Amphibolite, to the right of the blue dashed line, contains what is interpreted as the relict core of an unrecrystallized plagioclase phenocryst, to the left of the

Figure 25.1 (Continued) red dashed line. The margin of the phenocryst core is increasingly well-recrystallized toward the amphibolite, between the two lines. The relict core is a single crystal about 1 cm across that contains hornblende grains that may be part of recrystallized melt inclusions. The plagioclase core also contains fine-grained Fe-Ti oxide dust, indicated by grayish cloudiness in the plagioclase, visible in the lower left of C. Such dust is precipitated from Fe and Ti in the high-temperature plagioclase solid solution (e.g., Figs. 9.15–9.17). Exsolved Fe-Ti oxide dust is not found in metamorphic plagioclase in the vicinity, except rarely in tonalitic partial melt segregations at higher grade (Fig. 21.4). E, F) The chilled margin of a gabbroic pluton that survived metamorphism to granulite facies conditions. E) Contact between garnet-rich, aluminous gneiss host rock to the left of the red line, and the fine-grained gabbro, now garnet – pyroxene granulite, to the right. A pure plagioclase selvage occurs between. F) The same gabbro as in E, but from 75 cm into the pluton from the contact, where grain sizes are about 3 mm (Fig. 25.19). The granulite facies assemblages are apparent in F, where orthopyroxene – clinopyroxene – garnet coronas surround relict igneous olivine, and plagioclase cores are dark from abundant spinel inclusions (see Figs. 21.1C and D). Dresden Station, eastern Adirondacks, New York, USA. Abbreviations: E, epidote; EA, epidote and albite; G, garnet; H, hornblende; P, plagioclase; PP, plagioclase and pyroxene; Px, ortho- and clinopyroxene; Q, quartz; S, spinel.

Figure 25.2 Pillow lavas metamorphosed to epidote amphibolite facies conditions. The rocks here are almost undeformed at the scale of single pillows, even though they are nearly upside down. The pillows have smooth, rounded tops (toward the lower left) and deformed during extrusion to fill the spaces between underlying pillows (now toward the upper right). Thin selvages between the pillows are clearly different from the interiors, and shown in close-up in Figure 25.3. Løkken, Sør Trøndelag, Norway.

Figure 25.3 Close-up of the rims of two adjacent pillows in Figure 25.2. The chlorite- and magnetite-rich, black and dark-gray pillow rims (center) are mineralogically and texturally distinct from the epidote-rich pillow interiors (left and right). The rims represent the glassy crust that quenched on the pillow surfaces as they formed. During pre-metamorphic diagenesis, the glass probably devitrified, interacted with cold sea water, and changed chemical composition. The different chemical composition resulted in the rims having different metamorphic mineralogy and appearance. Løkken, Sør Trøndelag, Norway.

Figure 25.4 Deformed pillow lavas at epidote amphibolite facies. The pillows are stretched by a factor of about three, toward the upper-left and lower-right. The long white stripe (red arrow), and less regular white stripes and patches elsewhere, are crosscutting quartz veins. Figure 25.5 shows a close-up of one of the pillow margins. Soknedal, Sør Trøndelag, Norway.

Figure 25.5 Close-up of one of the pillow margins shown in Figure 25.4. One of the less-deformed pillows (left) is shown embedded a breccia of angular lava fragments (right), with light-colored material filling in between the fragments. The pillow itself contains small black pits, which are original vesicles from which carbonate filling has dissolved away. The presence of vesicles indicates that the lavas were not erupted in a great depth of water. White stripes in the pillow, to the left, are quartz-carbonate veins. Soknedal, Sør Trøndelag, Norway.

Figure 25.6 Deformed pillow lavas metamorphosed to epidote amphibolite facies conditions. The pillows themselves are outlined by smooth, dark-gray, anastomosing lines that separate pitted, lighter-gray, carbonate-bearing pillow interiors. Figure 25.7 shows a close-up of the pillows. Kvithyll, outer Trondheimsfjord, Sør Trøndelag, Norway.

Figure 25.7 Close-up of the pillow lavas in Figure 25.6. The dark pillow rims contain numerous red garnets, while the pillow interiors have no garnet but are rich in epidote, and a carbonate mineral that weathers out to give the pitted appearance. The glassy pillow rims underwent diagenetic chemical alteration as the quenched surface glass devitrified, resulting in the different mineralogy of pillow rims compared to cores. Kvithyll, outer Trondheimsfjord, Sør Trøndelag, Norway.

Figure 25.8 Fragmental rhyolitic volcanic rock, possibly pyroclastics, metamorphosed to greenschist facies conditions. With more deformation and at higher metamorphic grade, these might appear as thin, discontinuous layers, rather than recognizable angular volcanic fragments. The field width is about 80 cm. Duved, Jämtland, Sweden.

Figure 25.9 Rhyolitic pyroclastic rock, with light-colored clasts and darker, fine-grained material in between, metamorphosed to greenschist facies. The pen is oriented parallel to a spaced foliation and cleavage that wraps around the larger fragments. The fragments have resisted deformation and still have igneous textures, including recognizable, though sparse, phenocrysts. The dark matrix is made up mostly of a fine-grained, foliated mixture of white micas and quartz. Littleton, New Hampshire, USA.

Figure 25.10 Fine-grained dioritic dike cutting folded, layered tonalitic gneiss, all metamorphosed to upper amphibolite facies conditions. Some of the white layers are partial melts that were derived from the gneiss, but do not cut the dioritic dike. That indicates that the dike is younger than any layers in the gneiss. The dike itself has been weakly folded, and a thin, white pegmatite intruded along the right-hand dike margin. Quabbin Reservoir, Massachusetts, USA.

Figure 25.11 The dark-gray layer extending from lower left to upper right is a highly deformed basaltic dike that cuts layered quartzite, all metamorphosed to epidote amphibolite facies conditions. In essentially undeformed parts of this same quartzite unit, in Sweden, the dikes cut layering at about right angles. Along the top of the dike seen here, dark layers in the quartzite truncate against the dike, with intersection angles of about 1° (yellow arrows indicate truncations). Below the dike, between the blue arrows, are isoclinal fold hinges in the quartzite layering, with the fold opening to the upper right. Engan quarry, Oppdal, Sør Trøndelag, Norway.

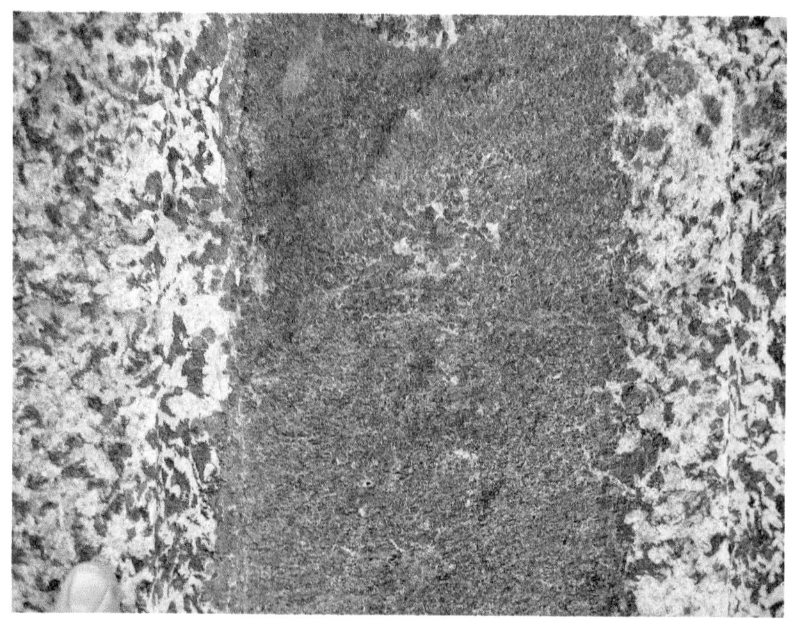

Figure 25.12 Ferrobasaltic dike cutting gabbroic anorthosite, both of which have been metamorphosed to granulite facies conditions. The dike has been converted to an assemblage containing about 50% garnet, along with orthopyroxene, augite, plagioclase, and Fe-Ti oxides. The surrounding gabbroic anorthosite still has a recognizably igneous texture, indicating that these rocks have not been deformed very much. North River, Adirondacks, New York, USA.

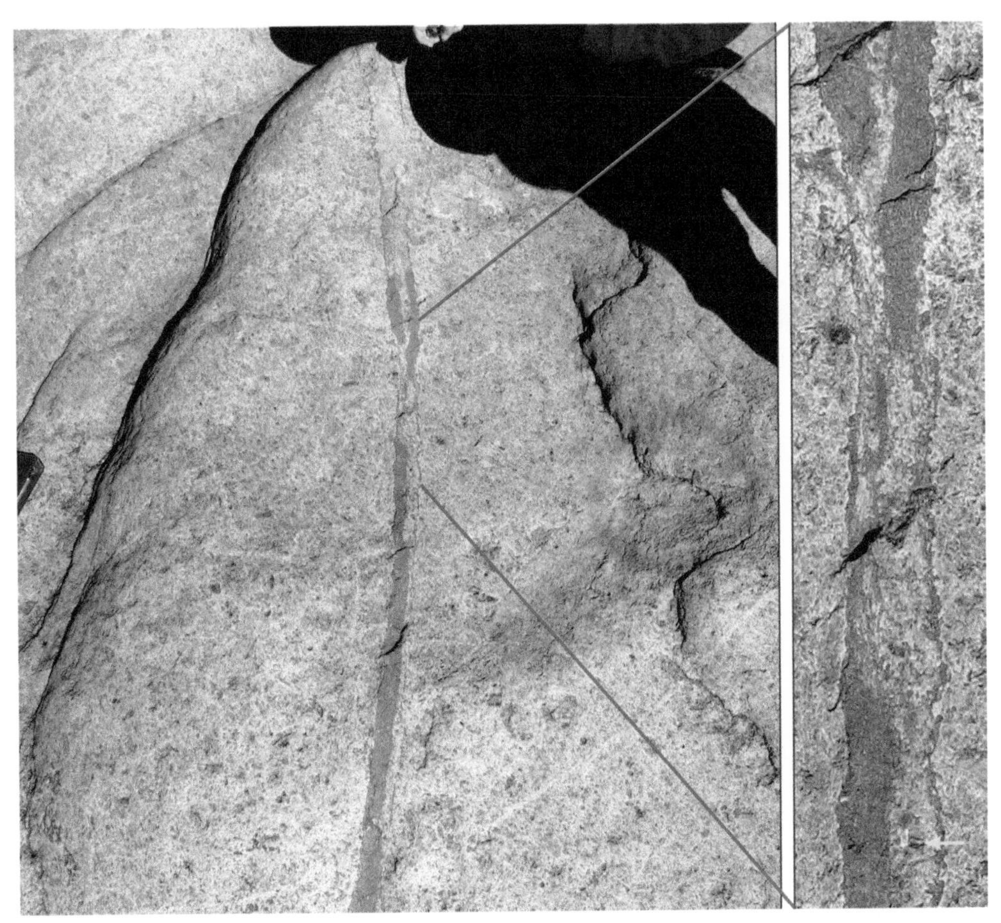

Figure 25.13 A small composite dike that cuts partially recrystallized anorthosite, both metamorphosed to granulite facies conditions but almost undeformed. The first liquid to intrude the solid anorthosite was ferrobasalt, now metamorphosed to a brown, fine-grained, garnet – two pyroxene – plagioclase granulite. The ferrobasaltic dike was intruded, quite literally, moments later by quartz syenite liquid. Although the larger feldspars in the syenite are gray (yellow arrow), they are perthite, not labradorite as in the anorthosite. The ferrobasalt and syenite magma apparently partially mixed in parts of the dike (enlarged view, right) indicating that the ferrobasalt magma was still partly liquid at the time syenite magma intruded. The fuzzy blue circle to the right in the larger image outlines an unrecrystallized 12 cm plagioclase porphyroclast in the anorthosite. Elizabethtown, Adirondacks, New York, USA.

Figure 25.14 Tonalitic gneiss, above and below the blue dashed lines, was cut by a composite dike, and later metamorphosed to epidote amphibolite facies conditions. The composite dike has three parts, labeled, of approximately equal thickness. Part 1 is a dark-gray amphibolite having sparse garnets and small, elongate white lenses of plagioclase. Compositionally it is andesitic (Hollocher, unpublished data). Part 2 is a medium-gray, relatively garnet-rich amphibolite with more plagioclase and less hornblende than Part 1. Part 2 is also andesitic, but richer in CaO and poorer in SiO_2 and Na_2O than Part 1. Part 3 is a garnet-free, fissile-weathering amphibolite that has a basaltic composition. It has much less SiO_2 and more MgO than the other parts. Part 3 has been boudinaged, with a quartz-filled inter-boudin space indicated by the red arrow. The intrusion sequence is not clear, but earlier dike parts were apparently solid at the time later parts were emplaced. Almvikneset, outer Trondheimsfjord, Sør Trøndelag, Norway.

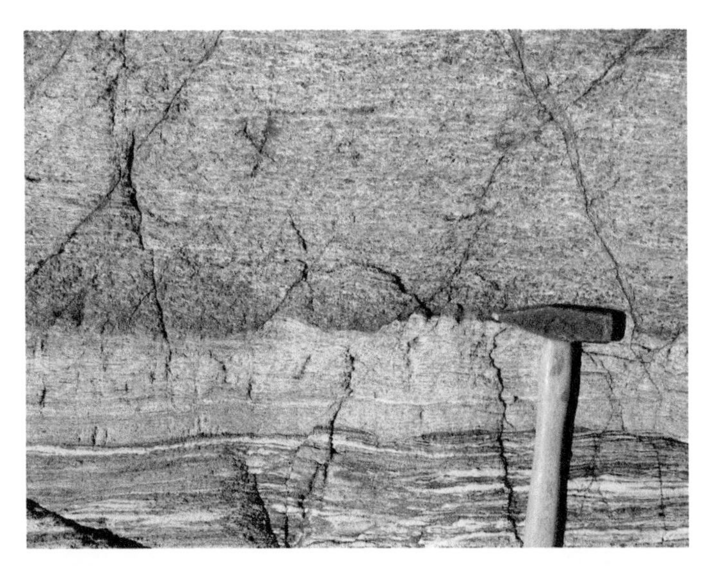

Figure 25.15 Modally graded layer in cumulate gabbro, metamorphosed to lower amphibolite facies conditions. Based on similar modally graded layers in gabbroic plutons, the grading can be interpreted to be right side-up, assuming the mafic-rich part of the graded layer is on the layer bottom. The rock is highly deformed, as indicated by a weak foliation in the upper layer, parallel to the hammer head, and the disrupted lowermost layer. The layers are cut by several small, late, normal faults, most of which are down-dropped to the left. Helland, Moldefjord, Møre og Romsdal, Norway.

Figure 25.16 This is part of a gabbroic pluton that retains well-developed, rhythmic, modally graded cumulate layering, despite having experienced granulite facies metamorphic conditions. The layers have been somewhat folded, but it remains beautifully intact. Haramsøy, Nordøyane, Møre og Romsdal, Norway.

Figure 25.17 This is a close-up of a basaltic boudin, metamorphosed to epidote amphibolite facies conditions. This rock has white, generally lath-shaped spots that physically look like plagioclase phenocrysts. In thin section (Figs. 25.1A, B) they also look like euhedral phenocrysts, but they have been replaced by a mixture of albite and clinozoisite. The phenocryst texture probably remained intact in this rock because the boudin is enclosed by quartzite, which was more ductile than the amphibolite during metamorphism. Dryna River, Oppland, Sør Trøndelag, Norway.

Figure 25.18 Metamorphosed and highly deformed basaltic dike (to the lower left of the yellow dashed line) cutting layering in granitic gneiss (center and right). The gneiss contains a large proportion of deformed microcline phenocrysts, now porphyroclasts, up to eight centimeters across, best seen in the lower-right part of the image. Because the dike cuts layering in the gneiss, the host rock had already been deformed at the time the dike intruded. Kjøra, outer Trondheimsfjord, Sør Trøndelag, Norway.

Figure 25.19 This shows a chilled margin at the contact between a metamorphosed gabbro body, which extends from the contact at the T-end of the arrow, to the upper right. Below the contact is the enclosing sillimanite – garnet gneiss. The gabbroic rock is now a garnet – two pyroxene granulite. Grain sizes are less than 0.1 mm at the contact (Fig. 25.1E), increasing to about 3 mm in the upper right part of this image, at the arrow head, about 75 cm away (Fig. 25.1F), and up to 1 cm for plagioclase crystals 15 m from the contact to the upper right. The rock breaks along smooth surfaces where it is fine-grained near the contact, and becomes progressively rougher to the upper right where the grains are larger. A chilled contact like this indicates that the pluton intruded at relatively shallow levels, where the surrounding rocks were cold. To preserve this texture the gabbro had to have remained undeformed through granulite facies metamorphism. A strong foliation in the enclosing gneiss, and foliated gneiss xenoliths in the gabbro, show that the gneiss had been metamorphosed prior to gabbro emplacement. Dresden Station, eastern Adirondacks, New York, USA.

Figure 25.20 This image shows two relatively extensive, coarse-grained plutonic bodies, metamorphosed to amphibolite facies conditions. To the right is a medium-gray hornblende tonalite gneiss, that has intruded into a light-gray biotite tonalite gneiss (left). The light-gray xenolith in the darker gray gneiss (top-right, red arrow) indicates the relative age of the two metamorphosed igneous bodies. Quabbin Reservoir, Massachusetts, USA.

Figure 25.21 Intrusive breccia metamorphosed to amphibolite facies conditions. The tonalitic host rock originally contained breccia fragments of dark, coarse diorite. Both have been deformed to the extent that the angular shapes of the original fragments are discernible only in less deformed regions, such as here. Elsewhere, they are exposure-spanning, somewhat discontinuous layers. Shelburne Falls, Massachusetts, USA.

Figure 25.22 Dark dioritic xenolith blocks engulfed in a light-gray tonalitic matrix, metamorphosed to amphibolite facies conditions. The xenoliths have been somewhat deformed, and are extended in a direction away from the photographer. Esvikneset, outer Trondheimsfjord, Sør Trøndelag, Norway.

References

Baltatzis, E. (1996) Blueschist-to-greenschist transition and the P-T path of prasinites from the Lavrion area, Greece. *Mineralogical Magazine, 60*, 551–561.

Bédard, J.H. (1999) Petrogenesis of boninites from the Betts Cove Ophiolite, Newfoundland, Canada: Identification of subducted source components. *Journal of Petrology, 40*, 1853–1889.

Bentley, C. (2009) Geology field trips as performance evaluations. Inquiry. *Journal of the Virginia Community Colleges, 14* (1), 77–93.

Blatt, H., Tracy, R.J., & Owens, B.E. (2006) *Petrology: Igneous, Sedimentary, and Metamorphic*, 3rd ed. New York, NY, Freeman and Co., 530.

Boucot, A.J., & Rumble, D. III. (1978) Devonian brachiopods from the sillimanite zone, Mount Moosilauke, New Hampshire. *Science, 201*, 348–349.

Boucot, A. J., & Thompson, J. B. Jr. (1963) Metamorphosed Silurian brachiopods from New Hampshire. *Geological Society of America Bulletin*, 74, 1313–1334.

Brown, M., Korhonen, F. J., & Siddoway, C. S. (2011) Organizing melt flow through the crust. *Elements*, 7, 261–266. doi: 10.2113/gselements.7.4.261

Cohen, K. M., Finney, S., & Gibbard, P. L. (2013) International chronostratigraphic chart: International Commission on Stratigraphy, 1. Accessed 2013 from: http://www.stratigraphy.org.

Connolly, J. A. D. (2009) The geodynamic equation of state: What and how. *Geochemistry, Geophysics, Geosystems*, 10, Q10014. doi: 10.1029/2009GC002540.

El-Shazly, A. K., Loehn, C., & Tracy, R. J. (2011) P-T-t evolution of granulite facies metamorphism and partial melting in the Winding Stair Gap, Central Blue Ridge, North Carolina, USA. *Journal of Metamorphic Geology*, 29, 753–780.

Elbert, D. C., Harris, A. G., & Denkler, K. E. (1988) Earliest Devonian conodonts from marbles of the Fitch Formation, Bernardston Nappe, north-central Massachusetts. *American Journal of Science, 288*, 684–700.

Eskola, P. (1922) On Contact Phenomena between Gneiss and Limestone in Western Massachusetts. *Journal of Geology, 20* (4), 265–394.

Evans, B. W. (1990). Phase relations of epidote-blueschists. *Lithos*, 25, 3–23.

Fowler, M. B., Williams, H. R., & Windley, B. F. (1981) The metasomatic development of zoned ultramafic balls from Fiskenaesset, West Greenland. *Mineralogical Magazine, 44*, 171–177.

Ghosh, S. K., Pandey, A. K., Pandey, P., Ray, Y., & Sinha, S. (2012) Soft-sediment deformation structures from the Paleoproterozoic Damtha Group of Garhwal Lesser Himalaya, India. *Sedimentary Geology, 261–262*, 76–89.

Goscombe, B. D., Passchier, C. W., & Hand, M. (2004) Boudinage classification: end-member boudin types and modified boudin structures. *Journal of Structural Geology, 26*, 739–763.

Grabau, A. W. (1920) *A Texbook of Geology, Part 1, General Geology*: Boston, MA, D. C. Heath and Co., Publishers, 864

Harwood, D. S., & Larson, R. R. (1969) Variations in the delta index of cordierite around the Cupsuptic pluton, west-central Maine. *American Mineralogist, 54*, 896–908.

Hitckcock, C. H. (1874) On Helderberg rocks in New Hampshire. *American Journal of Science, 3rd series, 7 (37–42)*, 468–476.

Hollocher, K. (2008) Origin of big garnets in amphibolites during high-grade metamorphism, Adirondacks, NY: Keck Geology Consortium, 2008 meeting at Smith College, Northampton, Massachusetts, USA, Symposium. *21*, 129–134.

Kaszuba, J. P., Williams, L. L., Janecky, D. R., Hollis, W. K., & Tsimpanogiannis, I. N. (2006) Immiscible CO_2-H_2O fluids in the shallow crust. *Geochemistry Geophysics Geosystems, 7* (10), 1–11. doi: 10.1029/2005GC001107

Lancaster, P. J., Fu, B., Page, F. Z., Kita, N. T., Bickford, M. E., Hill, B. M., McLelland, J. M., & Valley, J. W. (2009) Genesis of metapelitic migmatites in the Adirondack Mountains, New York. *Journal of Metamorphic Geology*, *27*, 41–54.

Leake, B. E., Woolley, A. R., Arps, C. E. S., Birch, W. D., Gilbert, M. C., Grice, J. D., Hawthrone, F. C., Kato, A., Kisch, H. J., Krivovischev, V. G., Linthout, K., Laird, J., Mandarino, J. A., Maresch, W. V., Nickel, E. H., Rock, N. M. S., Schumacher, J. C., Smith, D. C., Stephenson, N. C. N., Ungaretti, L., Whittaker, E. J. W., & Youzhi, G. (1997) Nomenclature of amphiboles: Report of the subcommittee on amphiboles on the International Mineralogical Association, Commission on new minerals and mineral names. *American Mineralogist*, *82*, 1019–1037.

Lien, T., Walker, R. G., & Martinsen, O. J. (2003) Turbidites in the Upper Carboniferous Ross Formation, western Ireland: reconstruction of a channel and spillover system. *Sedimentology*, 50, 113–148.

Marshak, S. (2012) *Earth: Portrait of a Planet*, 4th ed: New York, NY, W.W. Norton and Company, Inc., 819.

Matthews, D. H. (1971) Altered basalts from Swallow Bank, an abyssal hill in the NE Atlantic and from a nearby seamount. *Philosophical Transactions, Royal Society of London, Series A*, *268*, 551–571.

Morgan, E., & Hollocher, K. (2011) Big garnet rocks at Gore Mtn. & Warrensburg, NY: Geochemical evidence of fluid flow and conditions of garnet growth: *Geological Society of America, Abstracts with Programs*, *43* (1), 60.

Morton, P. S. (1985) *Tectonic Breccia of Metamorphosed Intrusive Igneous Rocks in an Acadian Shear Zone, Brooks Village, North-Central Massachusetts [M.S. Thesis]*: Contribution 54, Department of Geology and Geography, University of Massachusetts, Amherst, MA, 43 p.

Muñoz, M., Aguirre, L., Vergara, M., Demant, A., Fuentes, F., & Fock, A. (2010) Prehnite-pumpellyite facies metamorphism in the Cenozoic Abanico Formation, Andes of central Chile (33°50'S): chemical and scale controls on mineral assemblages, reaction progress and the equilibrium state. *Andean Geology*, *37*, no. 1, 54–77.

NAGM: North American Geologic-Map Data Model Science Language Technical Team (2004) Report on progress to develop a North American science-language standard for digital geologic-map databases, Appendix B: Classification of metamorphic and other composite-genesis rocks, including hydrothermally altered, impact-metamorphic, mylonitic, and cataclastic rocks. In: Soller, D.R., ed. *Digital Mapping Techniques, 2004 Workshop Proceedings*. U.S. Geological Survey Open-File Report 2004–1451. pp. 56

Oh, C., Liou, J. G., & Maruyama, S. (1991) Low-temperature eclogites and eclogitic schists in Mn-rich metabasites in Ward Creek, California; Mn and Fe effects on the transition between blueschist and eclogite. *Journal of Petrology*, *32*, 275–301.

Passchier, C. W., & Simpson, C. (1986) Porphyroclast systems as kinematic indicators. *Journal of Structural Geology*, 8, 831–843.

Peacock, S. M., & Goodge, J. W. (1995) Eclogite-facies metamorphism preserved in tectonic blocks from a lower crustal shear zone, central Transantarctic Mountains, Antarctica. *Lithos*, *36*, 1–13.

Poldervardt, A. (1953) Metamorphism of basaltic rocks: A review: *Geological Society of America Bulletin*, *64*, 59–274.

Robertson, S. (1999) *British Geological Survey Rock Classification Scheme, Volume 2: Classification of Metamorphic Rocks*. Research Report RR99-02, Nottingham, UK, British Geological Survey, 24.

Salters, V., & Stracke, A. (2004). Composition of the depleted mantle. *Geochemistry, Geophysics, Geosystems, 5* (5), 1–27. doi: 10.1029/2003GC000597

Schumacher, J.C. (1988) Stratigraphy and geochemistry of the Ammonoosuc Volcanics, central Massachusetts and southwestern New Hampshire. *American Journal of Science, 288*, 619–663.

Spear, F. S., & Selverstone, J. (1983), Water exsolution from quartz: Implications for the generation of retrograde metamorphic fluids. *Geology*, 11, 82–85.

Spry, A., & Burns, K. (1967) The origin of deformed conglomerate and pseudo deformed conglomerate; with particular reference to the rocks at Goat Island, Tasmania. *Papers and Proceedings of the Royal Society of Tasmania, 101*, 271–277.

Suchý, V., Šafanda, J., Sýkorová, I., Stejskal, M., Machovič, V., & Melka, K. (2004) Contact metamorphism of Silurian black shales by a basalt sill: geological evidence and thermal modeling in the Barrandian Basin: *Bulletin of Geosciences, Czech Geological Survey*, 79 (3), 133–145.

Thompson, P. J. (1988) Trip C-1, Geology of Mount Monadnock. In: Bothner, W. A., ed., *Guidebook for Field Trips in Southwestern New Hampshire, Southeastern Vermont, and North-Central Massachusetts*. Keene, New Hampshire, New England Intercollegiate Geological Conference, 80th annual meeting. pp. 268–273.

Tracy, R. J. (1978) High grade metamorphic reactions and partial melting in pelitic schist, west-central Massachusetts. *American Journal of Science*, 278, 150–178.

Tracy, R. J., Robinson, P., & Thompson, A. B. (1976) Garnet composition and zoning in the determination of temperature and pressure of metamorphism, central Massachusetts. *American Mineralogist, 61*, 762–775.

Tsai, C.-H., Iizuka, Y., & Ernst, W. G. (2013) Diverse mineral compositions, textures, and metamorphic P–T conditions of the glaucophane-bearing rocks in the Tamayen mélange, Yuli belt, eastern Taiwan. *Journal of Asian Earth Sciences, 63*, 218–233.

Vrijmoed, J. C., Van Roermund, H. L. M., & Davies, G. (2006) Evidence for diamond-grade UHP metamorphism and fluid interaction in the Svartberget Fe-Ti garnet peridotite/websterite body, Western Gneiss Region, Norway: *Mineralogy and Petrology, 88*, 381–405.

Westphal, M., Schumacher, J. C., & Boschert, S. (2003) High-temperature metamorphism and the role of magmatic heat sources at the Rogaland Anorthosite Complex in southwestern Norway. *Journal of Petrology, 44*, 1145–1162.

Wheeler, B. D. (1999) Analysis of limestones and dolomites by X-ray fluorescence. *The Rigaku Journal*, 16 (1), 16–25.

Winter, J. D. (2001) *An Introduction to igneous and metamorphic petrology*. Upper Saddle River, New Jersey, USA, Prentice-Hall, Inc., 697.

Wintsch, R. P., Aleinikoff, J. N., Walsh, G. J., Bothner, W. A., Hussey, A. M., II, & Fanning, C. M. (2007) Shrimp U-Pb evidence for Late Silurian age of metasedimentary rocks in the Merrimack and Putnam-Nashoba terranes, eastern New England: *American Journal of Science, 307*, 119–167.

Zen, E-an, ed. (1983) *Bedrock Geologic Map of Massachusetts*: U.S. Geological Survey, Washington, D.C., 1:250,000. 3 sheets.

Glossary of Terms, Rock Types, and Minerals

GLOSSARY OF TERMS

Terms are defined as they are used in this book. In other contexts some terms may also have other meanings.

Accretionary wedge The part of a subduction zone system nearest the trench, between the surface and the subducting oceanic plate. It is made of scraped-up ocean floor and trench sediment, commonly with entrained fragments of volcanic and ultramafic rocks. The deeper parts of accretionary wedges, and also the subducting oceanic plate, are where blueschists and eclogites are formed.

Activity The effective concentration of chemical components that take part in chemical reactions, used in thermodynamic calculations.

Advection Flow of fluid by movement down pressure gradients, rather than by diffusion.

Alkali chemical component Rich in the alkali elements Na and/or K.

Allochthon A large mass of rock that has been transported a substantial horizontal distance along a fault, and is therefore very much out of place compared to underlying rocks.

Alteration The modification of a volume of rock by fluids, generally with conversion of a higher temperature or volatile-poor assemblage to a lower temperature, volatile-rich assemblage. For example, partial replacement of feldspars in a rock by fine-grained white mica (sericite).

Anastomosing The irregular weaving and crossing of features like foliation or fractures, something like braided hair or braided rivers.

Anhedral Single crystals (one continuous lattice) that have no clear external crystal shape, such as flat facets.

Anhydrous No water (as molecular H_2O or OH^-) in a rock or a mineral crystal lattice.

Anisotropic A rock or crystal that is different in some directions than others. Anisotropic rocks have a foliation or lineation (e.g., not a granulite), anisotropic minerals include all but those in the isometric (cubic) crystal system.

Annealed A deformed rock that remained hot for long enough time that the grains have recrystallized to reduce or eliminate strain in the crystals. Strained crystals may be bent, or may contain sub-domains of slightly different crystallographic orientations. In fully annealed rocks, most mineral grains are relatively large and have approximately 120° grain boundary intersections.

Anoxic Without oxygen, generally referring to fresh or salt water where bacterial respiration has used up all the free oxygen (O_2; bogs, deep, stagnant fjords, high-productivity coastal waters, etc.). In such environments, sediments tend to accumulate organic carbon, sulfides, and some redox-sensitive trace elements like V and Mo.

Anticline In less-deformed regions this refers to a fold arched (convex) upwards. In more deformed regions, with overturned folds, the word is restricted to those having progressively younger rocks on the convex side.

Antiform A convex-upward fold in which either the relative age of the different layers is unknown, or in which the layer age sequence varies because of folding or faulting.

Aqueous fluid Referring to H_2O-rich fluid.

Arc A volcanic arc, the plate tectonic structure that includes a chain of volcanoes above a subducting oceanic plate.

Archean Geologic Eon for rocks older than 2500 million years.

Assemblage If the context is a group of minerals, the word refers to the group of phases coexisting in a rock, which may include a fluid phase during metamorphism. If the context is rocks, then it refers to a set of rocks in close proximity that formed, or came together, in a similar or related way. Examples are a series of thrust slices making tectonostratigraphic package, the set of rocks making up a mélange, and the texturally and mineralogically distinct rock types of a migmatite.

Augen Eye-shaped porphyroclasts.

Aureole The contact metamorphosed, metasomatized, or altered region around the heat or fluid source.

Axial surface The surface that connects the hinges of a series of nested folds, approximately bisecting the angle between the fold limbs.

Bar Unit of pressure, corresponding to 100,000 Pascals, approximately 1 atmosphere pressure at sea level, 1 kg/cm^2.

Basin A geographic region that is (or was) subsiding to accumulate a large thickness of sediment.

Bed A distinct layer of sedimentary or volcanic rock.

Binary mixture A mixture of two things, such as two different magmas or sediment types.

Boudin This refers to tablular (flat) bodies of strong rock, surrounded by more ductile rock, that has been stretched and broken into long, sausage-like segments.

Breakdown The reduction in amount, or loss, of a mineral reactant during a chemical reaction. The lost minerals are replaced by new minerals, the reaction products.

Brittle The ability to break into angular fragments.

Calcareous Contains carbonate minerals, like calcite or dolomite.

Calcic Particularly calcium-rich.

Cataclastic texture A fault rock composed of broken, angular fragments. These form from faulting of brittle rocks.

Cement The mineral material that precipitated between grains in a sedimentary rock that holds the grains together. The cement can be externally derived, such as a calcite-cemented quartz sandstone. It can also be derived from the minerals in the sediment itself, as in quartz-cemented quartz sandstone where the silica for the cement came from the grains themselves where they touch.

Channelways The open spaces, larger than crystal grain boundaries, through which fluids or magmas flow.

Chilled margin The fine-grained margin of a pluton that quickly cooled against the cold rocks at its contact.

Clast A broken fragment of rock, such as a sand grain, pebble, or angular fragment in a fault zone.

Cleavage The property of a rock or mineral to break more easily in particular directions.

Coarsen To become coarser. This may result from grain growth during recrystallization, or growth of a mineral during reactions that produce it.

Competent rock A rock body that is stronger than adjacent rock, with the ability to deform in a less ductile way.

Conchoidal fracture To break like glass, with curved fracture surfaces. Characteristic of glassy rocks like obsidian, some extremely fine-grained, hard rocks like chert, or minerals like quartz that lack cleavage.

Concretion A volume of a sediment or sedimentary rock that is distinctive from surrounding material because of the intragranular cement in that part. For example, calcite cement can produce concretions in otherwise un-cemented silt, or siderite can form concretions in otherwise self-cementing shale. The concretion is thus made of both the distinctive cement, and the enclosed grains, in a small rock volume. Metamorphosed concretions can be mineralogically distinct from adjacent rock.

Consumed Minerals that decrease in abundance, or disappear, during progress of a chemical reaction.

Contact The surface (or line on a map or outcrop) along which two rock types or geologic units touch.

Contact metamorphism Metamorphism that is restricted to a limited zone around a particular heat source, such as a pluton or a hot slab of rock from the deep crust.

Continental Sedimentary material derived from continental rock. Alternatively, material deposited above sea level, exposed to air, as implied by the presence of depositional features, or abundant oxidized iron. May also refer to continental crust in general.

Convection The gravitationally-induced movement of material caused by density differences that result from differences in temperature or composition. Hydrothermal systems, large volumes of magma, and cooling cups of coffee can convect.

Corona A mineral garland, ring, moat, or halo, or nested sequences of the same, around some center mineral or rock. These generally indicate a series of metamorphic reactions that were limited by the diffusion of the chemical components needed for the reactions.

Crenulations A series of small-scale, parallel folds that are related to displacements along spaced cleavage surfaces.

Cretaceous Last geologic period in the Mesozoic Era, extending from about 145 to 66 million years ago.

Crossbed Minor beds that are deposited at an angle to the major beds in which the crossbeds are a part. Crossbedded sandstone is the best example.

Crosscut contact Contacts are the surfaces along which two rock types or rock units touch. Erosion, faulting, and intrusion of magma are processes that can cut (truncate) those contacts. The process that did the cutting is younger than the rocks that were cut.

Decarbonation To produce CO_2 fluid from the decomposition of carbonate minerals.

Decompress To bring material from high-pressure to low-pressure (great depths to shallow depths).

Deformation To change the shape of rock, caused by differential stresses (applied forces) causing strain (shape change).

Dehydration To produce H_2O fluid from the decomposition of minerals that contain OH^- groups or molecular H_2O.

Detachment fault A large-scale, generally low-angle normal fault.

Detrital Solid material, transported and deposited as sediment. Sand, clay, and pebbles are detrital materials.

Devitrified To have turned crystalline from a glassy (vitreous) state.

Devonian Geologic period in the Paleozoic Era, between about 419 and 359 million years ago.

Dextral shear Picturing yourself standing on one side of a shear zone, looking across it, the opposite side moved to the right.

Diagenesis Processes that occur to change sediments after deposition, or sedimentary rocks while they are still in the sedimentary rock realm. Such processes include cementing grains, changes to clay minerals, recrystallization of fine-grained calcite, and so on.

Differentiation To separate one thing from another, such as to separate silicate liquid from restite in partially melted rock, or crystals separating from a cooling body of magma.

Diffusion The thermally-induced movement of molecules, chemical elements, or heat through a material by atomic-scale, random walk-type motion.

Dike A plutonic rock body that is thin, flat, and crosscuts local layering. It is produced by magma filling a fracture.

Dip The downhill direction on a tilted surface, such as a cleavage surface. The dip direction is perpendicular to the strike line on the same surface.

Dismembered Broken up into pieces.

Ductile Ability to deform without fracture, like wet clay or bread dough.

Enclaves Unusual objects in a host of different material, like blocks of basaltic rock in granite, or an unusual mineral assemblage isolated inside of another mineral.

Endothermic A chemical reaction or other process that absorbs heat.

Equant An object of approximately equal dimensions in all directions, such as minerals in the isometric (cubic) crystal system, like garnet.

Equilibrium A thermodynamic term meaning that a system of phases (minerals, fluids) has the lowest possible energy. No reactions can take place in a system at equilibrium.

Euhedral Crystals with well-formed facets.

Eutectic The lowest temperature at which a group of phases will melt. More specifically, the temperature at which a set of minerals melts to liquid of the same net composition.

Exhausted A reactant that vanishes during a chemical reaction, ending further progress of that reaction.

Exhumation The processes that bring deep rocks to the surface. These processes include erosional removal of material, tectonic removal of material along detachment faults, and tectonic escape of material from depth along thrust faults.

Exothermic A reaction or other processes that releases heat.

Exsolution The unmixing of two phases from an original homogeneous solid solution. For example, the unmixing of a high-temperature K-Na feldspar into thinly interlayered K-rich and Na-rich feldspar at lower temperature.

Extension The movement of two points in a rock farther from one another. This may result from actual tension at shallow depths, but at greater depth it results from less compressive stress in the extension direction compared to the other directions.

Extrude To undergo ductile deformation by extending along a one axis while contracting in the other two.

Fabric The arrangement of structural elements in a rock, such as foliations, lineations, and fold axes.

Facies A broad pressure-temperature range defined by a particular mineral assemblage in common rocks, such as amphibolite facies being defined by the hornblende-plagioclase mineral assemblage in common basaltic rocks. The term may have specialized meanings, for example, the sericite-pyrite facies of a particular ore deposit.

Fault A surface or relatively thin zone in rock along which one side has moved relative to the other.

Felsic A mineral that is rich in alkali elements (e.g., Na, K) and silica (SiO_2), and poor in Mg, Fe, and Ca, and generally light in color. Quartz, albite, and muscovite are examples of felsic minerals. The term also refers to rocks that are rich in felsic minerals. Granite, granodiorite, and tonalite are examples of felsic rocks.

Fissile To easily break into small, platy fragments.

Flattening To undergo ductile deformation by contracting (flattening) along one axis and extending along the other two.

Fluid Refers to a low-density material with no shear strength: a gas, liquid, or supercritical fluid, typically dominated by H_2O or CO_2.

Flute cast A small erosional scour, typically 1-10 cm across and 10-100 cm long, produced by turbulent flow of a turbidity current over easily eroded, underlying sediment (the scour is the mold). The scour is then filled with sandy sediment deposited from the turbidity current (the filled mold is the cast). Where there are many of them together, they can slightly resemble the grooves, or flutes, on Roman columns.

Fluvial deposits Sedimentary deposits from a river or stream.

Flux Mass flow of a fluid or other material or chemical component in one direction. Alternatively, a material that, added to a rock, promotes melting by lowering the melting temperature to below the local temperature. H_2O is the most common such flux.

Fold Layers or other once-planar features (sedimentary bedding, dikes, cleavage surfaces, geologic unit contacts, etc.) that have been distinctively bent into a curved shape.

Fold axis A line along the fold hinge, connecting points of maximum curvature.

Fold hinge The part of a folded feature where it is most sharply curved.

Fold limb The less-curved parts of folded layers that are between fold hinges.

Foliation The parallel-planar arrangement of features in a rock that result from deformation. These may be layers (commonly called layering), parallel-aligned platy minerals like muscovite, or flattened clasts in a deformed conglomerate, for example.

Forearc The part of a subduction zone system between the volcano chain and the trench, and above the subducting oceanic plate. It includes the accretionary wedge, possibly old ocean floor and other parts of the arc that formed when subduction began, and overlying sedimentary and volcanic rocks.

Formation A formally named rock unit, such as the Littleton Formation in northeastern USA, the Cauê Formation of southeastern Brazil, or the Rongbuk Formation at the base of Mt. Everest in Nepal.

Fragmental Made of angular fragments of rock, minerals, or fossils.

Fresh A rock or mineral in its original state, not weathered, hydrothermally altered, or retrograded, depending on the context.

Geotherm The curve of temperature vs. pressure or depth in the Earth.

Geothermal Heat within the earth, generally referring geothermal gradient or to geothermally heated, fluid-rich systems.

Graben A block of rock that has moved downward along normal faults on either side of the block.

Graded bed A layer that varies regularly in some way from bottom to top, such as grain size or mineral proportions. Examples are turbidite deposits (grain size and the proportions of sand, silt, and clay), and igneous cumulate layers (igneous mineral proportions).

Grain Ideally a single crystal, with the same, continuous lattice from one side to the other. More realistically, it need only look like a singe crystal by whatever analytical technique one is using. Bare eyes and hand lenses tend to find large grains. Optical microscopes and more sophisticated techniques tend to find smaller grains (see subgrains).

Horizon A continuous surface that can be traced through an outcrop or across a landscape, for example a lithologically distinctive part of a formation, a contact between formations, or a surface of constant age.

Hydrated Has had water added to it, in the form of H_2O molecules or OH^- groups, such as a volume of retrograded or hydrothermally altered rock.

Hydraulic fracturing Rocks may contain an interconnected network of fluid or silicate liquid at grain boundaries or other spaces. Because fluids have densities lower than the surrounding rock, rock pressure at the bottom of the network will be higher than the fluid pressure, and at the top of the network the fluid pressure will be higher than the rock pressure. If the pressure difference at the top exceeds rock strength, the fluid will fracture the rock and migrate upwards. At the same time, the space occupied by fluid at the bottom of the network will close up.

Hydrothermal Related to hot water, as in a hydrothermally altered rock or hydrothermal system.

Hydrous Has water in it, in the form of H_2O molecules or OH^- groups.

Igneous Formed from molten rock.

Inclusion One mineral, or bit of fluid (a fluid inclusion), completely enclosed in another mineral.

Incompatible element Elements like Th, Cs, and La that tend to be incompatible with the crystal structures of common minerals in igneous rocks, and so tend to segregate into the surrounding magmatic liquid.

Indicator mineral A mineral that is characteristic of a particular metamorphic grade, facies, or zone, in a metamorphosed region, an ore deposit, or a hydrothermal field.

Interbedded Two or more different sedimentary, or volcanic rock types, deposited in alternating sequence.

Intrusion One rock emplaced into another, like an basaltic dike or sill, or ductile marble that flowed into a fracture.

Invading Refers to the movement of one material into another, either displacing it (e.g., magma intruding fractures, pushing rock aside), or not (e.g., fluid flowing into a dry rock along grain boundaries).

Isoclinal A fold where the limbs are parallel, or nearly so.

Isotropic Refers to materials having the same properties in all directions, such as rocks that lack lineation, foliation, or other fabric elements.

Joints Sets of parallel-planar or gently curved fractures in brittle rocks, caused by tectonic or other stresses within the rock.

Jurassic Geologic period in the Mesozoic Eon, between about 201 and 145 million years ago.

Kinetics The science and mathematics of reaction rates. At any given condition, some reactions are fast, such as the transformation of beta quartz to alpha quartz with decreasing temperature, and some are slow, such as the transformation of kyanite to sillimanite with increasing temperature.

Lamellae Thin interlayering of one mineral with another, such as forms during unmixing (exsolution) of a solid solution into two phases. In perthite, a high-temperature K-Na feldspar solid solution has unmixed at lower temperature to form alternating thin layers of K-rich and Na-rich feldspar.

Leucocratic Rocks at the light-colored end of the spectrum, implying abundant quartz, feldspar, or other light-colored minerals, and relatively little biotite, hornblende, or other dark-colored minerals.

Leucosome The light-colored rock in a migmatite that is usually interpreted to be crystallized, segregated partial melt.

Lineation The parallel alignment of line-like elements in a rock. These may include elongate minerals, elongated clasts, trails of disrupted mineral grains, the intersection lines of two planar features like cleavage sets, and so on. Analogous to combed, straight hair (where the lineation curves from nearly horizontal to vertical), or wheat stalks standing up in a field (where the lineation is oriented approximately vertically).

Lithification The processes that turns loose materials like sand or volcanic ash to rock, such as by precipitation of intragranular cement.

Lithosphere The relatively cold and rigid crust and upper part of the mantle that make up the tectonic plates. Lithosphere thickness ranges from 10 kilometers or so at mid-ocean ridge axes up to 300 km under ancient continental shields.

Lithostatic pressure Pressure produced by the weight of overlying rock.

Mafic Mg- and Fe-rich minerals, such as olivine, enstatite, hornblende, or biotite. Also refers to mafic rocks, which are rich in mafic minerals and so tend to be dark-colored.

Magma Liquid rock, or liquid rock plus crystals, below ground.

Magma chamber A substantial volume of magma within the solid Earth. These can be storage reservoirs for volcanoes, and can crystallize to form plutonic rocks.

Mélange Generally refers to a severely deformed, large volume of rock that contains abundant fragments of diverse rock types, such as ultramafic rocks, pillow lavas, sandstones, and blueschist. Many mélanges are interpreted to be parts of subduction zone accretionary wedges.

Melanosome The darkest part of a migmatite. These are characteristically at the margin between the leucosome (light-colored, segregated partial melt) and the surrounding rock.

Melt Liquid from melting rock, with or without crystals.

Mesosome The intermediate-darkness part of a migmatite. This may be interpreted as the rock from which melt was extracted (restite), also referred to as the neosome.

Metamorphic belt An enormous elongate region in which metamorphism occurred a long time ago, that is now exposed at the surface. These are generally the exposed interiors of former mountain ranges. They are elongate because they usually form along subduction zones, or arc collision or continental collision zones. Metamorphic belts may also form in continental rift zones and along hot spot tracks.

Metamorphic differentiation The segregation of materials on a local scale during deformation, such as during the development of a spaced cleavage set.

Metamorphic field gradient Metamorphic conditions as indicated by the dominant, prograde metamorphic mineral assemblages in rocks across a regional metamorphic belt.

Metamorphic grade A pressure-temperature region that is defined by characteristic mineral assemblages in certain compositionally uniform metamorphosed rocks. For example, pelitic rocks metamorphosed to the medium-pressure biotite grade probably have biotite and chlorite, whereas similar rocks at lower grade might have chlorite only, and rocks at higher grade might have garnet in addition to biotite.

Metamorphism The change of a precursor rock (the protolith) under conditions of increased temperature (typically above 200°C), pressure, deformation, or chemical flux. Compared to protoliths, metamorphic rocks have generally changed shape, mineralogy, chemical composition, or some combination.

Metasomatism The movement of chemical components into or out of (or both) a rock volume, typically but not necessarily involving movement of a fluid phase.

Metastable A mineral that is thermodynamically unstable, but present anyway because the reactions that should have destroyed it (to make a stable assemblage) were too slow. Diamond at Earth's surface is metastable, for example.

Microlithon The relatively undeformed rock between spaced cleavage surfaces.

Migmatite Mixed rock, made up of an igneous-appearing part (the leucosome, usually interpreted to be segregated partial melt) and the metamorphic-appearing part (mesosome, melanosome).

Mineralization An odd term, perhaps because most of the world is made of minerals. This specifically refers to the addition of chemical components to a volume of rock, with precipitation of new minerals, especially economically valuable ore minerals.

Modally graded Regular variations in mineral proportions (the mode) from one place to another. Some igneous cumulate layers and turbidity current deposits are modally graded, and visible grading can survive metamorphism even if the original minerals and grain sizes do not.

Modes Volume or weight proportions of minerals in a rock.

Mullions Lineations defined by elongate boudins or parallel, thickened fold hinges, over which the cleavage is contorted.

Normal fault An inclined fault where the upper block moved down relative to the lower block.

Nucleation The initial process that ultimately grows into a crystal, a fold, a fracture, or some other feature. For example, an initial, minute chance arrangement of atoms in a supersaturated system can form a stable crystal nucleus. Atoms are added to the nucleus, eventually growing into a sizable crystal.

Oikocryst A large mineral grain that contains a great many inclusions of one or more other minerals. It usually indicates that the larger grain overgrew the others. This term usually refers to crystals that have grown from interstitial magmatic liquid in an igneous cumulate.

Ophiolite A sequence of rocks, typically kilometers thick, with the sequence: ultramafic rocks at the bottom, basaltic rocks including pillow lavas in the middle, and deep marine sediments at the top. These are generally interpreted to be fragments of oceanic crust in one form or another. Historically, they were thought to have been made of crust formed at mid-ocean ridges. Most, however, are now thought to be forearc crust, or crust of another part of the arc system.

Ordovician Geologic period in the Paleozoic Era, between about 485 and 443 million years ago.

Orogen A mountain belt, either currently mountainous (e.g., the Himalayan orogen) or ancient and eroded nearly flat (e.g., Pan-African orogen).

Outcrop An exposure of bedrock on Earth's surface. The term does not apply to boulders, even large, mostly buried ones, or exposures on the insides of tunnels or caves.

Overturned fold A fold in which one or both limbs have been deformed to being past the vertical.

P Pressure.

Penetrative cleavage A cleavage that is spaced at the scale of individual mineral grains, thus permeating, or penetrating, the whole rock volume (except perhaps for some porphyroblasts).

Peraluminous A rock having an amount of Al in excess of that needed to make feldspars from Ca, Na, and K. This permits left-over Al to enter Al-rich minerals such as muscovite, aluminosilicates, and cordierite.

Petrography The systematic study and classification of rocks, particularly with regard to mineralogy, textures, and chemical compositions.

Petrology The study of rocks at small scale, involving minerals, mineral compositions, the physical and chemical relationships between rocks and minerals, and rock fabrics.

Phanerozoic Geologic eon extending from about 541 million years ago to the present. Encompasses the fossil record of organisms with hard parts.

Phase A physically separatable, distinct part of a material system. A low-temperature rock containing chlorite, talc, antigorite, magnesite, H_2O-rich liquid, CO_2-rich liquid, and H_2O-CO_2 gas has seven phases: four solid phases, two immiscible liquid phases, and one gas phase.

Phenocrysts Large igneous crystals that are set in a much finer-grained matrix. They indicate a two-stage crystallization history: early slow crystal growth that allowed a few, large crystals to form, followed by rapid crystallization (perhaps faster cooling closer to the surface, or after magmatic volatile loss), that caused many fine-grained crystals to grow.

Picritic basalt An unusually Mg-rich basalt, with or without olivine phenocrysts (usually with).

Pillow lava Lava under water very quickly forms a frozen, hard crust. The flow front advances as magma pressure within the flow breaks the crust, extruding lava through the breaks like toothpaste from a tube. Cross sections of the extruded lava, on an outcrop surface, look rather like pillows.

Plate tectonics Theory of the large-scale geologic structure and movement of large sections (plates) of the relatively cold, rigid, outer 10-300 km of the Earth (the lithosphere). The plates include the crust and uppermost mantle.

Play of colors Here this refers to a diffraction effect as light is scattered off of closely-spaced exsolution lamellae in minerals like some plagioclase and gedrite. Light wave interference causes particular colors of light to be directed in specific directions.

Plunging fold A fold where the fold axis extends downward into the ground.

Pluton A body of rock formed from the crystallization of magma below ground. Although dikes and sills are plutons, in the field some workers reserve this term for larger or more irregular bodies.

Poikilitic A textural term referring to a mineral grain, such as a porphyroblast, with a large proportion of smaller mineral inclusions.

Porphyritic An igneous rock texture in which abundant large crystals (phenocrysts) are in a matrix of much smaller grains.

Porphyroblast A large crystal that has grown in the solid state in a metamorphic rock, that is set in a matrix of much smaller grains.

Porphyroclast A deformed crystal in metamorphic rock, such as an igneous phenocryst, metamorphic porphyroblast, or the dispersed parts of a dismembered pegmatite.

Porphyroclastic rock A rock with a lot of porphyroclasts.

Precambrian Geologic eon older than about 541 million years ago.

Prograde metamorphism Metamorphism that occurs during progressively increasing temperatures, and commonly also increasing pressures.

Protolith The precursor igneous or sedimentary rock that was metamorphosed to become a metamorphic rock. Identifying the protolith is critical for unraveling pre-metamorphic geology.

Pseudomorph A crystal that has been replaced by one or more other minerals, as by a metamorphic reaction, while retaining the shape of the original crystal.

Ptygmatic fold A folded competent layer in much more ductile rock, forming tight, irregular, dissimilar folds.

Recrystallization The in-place, progressive transformation of deformed or small grains into larger, more perfect ones.

Recumbent fold A fold having a nearly horizontal axial surface.

Reequilibrate The process of changing mineralogy or mineral compositions to approach equilibrium after a change in metamorphic conditions. This occurs through chemical reactions.

Refractory A material that has a high melting temperature (or at least higher than what it is being compared to).

Regional metamorphism Metamorphism that covers a broad region, not related to any particular pluton or other obvious heat source.

Restite The unmelted, solid material left behind after partial melting and melt removal.

Retrograde metamorphism The replacement of some or all minerals of a metamorphic rock prograde assemblage with those characteristic of lower temperature and/or pressure conditions. Typically, this will involve influx of fluids to make the retrograde minerals. An example is the replacement of staurolite and biotite by an intergrowth of a much more hydrous, lower-grade assemblage of chlorite and muscovite.

Satin A special fabric weave having a glossy surface. It is usually made of long-filament fibers like silk or polyester.

Schistosity A foliation and cleavage defined by abundant, coarse sheet silicates such as mica, chlorite, or talc.

Sediment A deposit on Earth's solid surface of material that was transported by wind, water, or flowing ice. Includes clastic sediment made of particles eroded from other rocks (e.g., sand, clay), and chemical sediments that were precipitated from water solution by organisms or inorganic processes (e.g., limestone).

Segregation The separation of one thing from another, and its concentration into a limited space. Examples include separation of partial melts into opening extensional fractures, and the concentration of micas along spaced cleavage surfaces.

Selvage A thin layer or coating at the edge of a rock body, such as a black coating of biotite on the margin of a migmatite leucosome.

Shear Refers to the deformation of rocks. Pure shear (flattening) involves compression in one direction and extension in the other two. Simple shear is like the parallel sliding of a deck of cards. Extrusion, a form of pure shear, refers to compression in two directions and extension in another. Reality for rocks ranges between these end member strain types.

Sheath fold A fold that has a shape like a knife blade or sword sheath, or fingers in a glove.

Silica SiO_2 as a chemical component, such as that dissolved in water or as part of an amphibole structure, rather than a phase like quartz.

Silicate A mineral with SiO_4^{4-} tetrahedra as important chemical and structural components, or a rock dominated by silicate minerals.

Silicate liquid Melted rock in which silica (SiO_2) is the major part.

Silurian A geologic period in the Paleozoic Era between about 443 and 419 million years ago.

Sinistral shear Picturing yourself standing on one side of a shear zone, looking across it, the opposite side moved to the left.

Slickensides Smooth or shiny fault surfaces, typically having striations parallel to the fault slip direction.

Spaced cleavage A cleavage in which the cleavage surfaces are separated from one another by some distance, typically a millimeter to several centimeters. The cleavage surfaces are separated by relatively undeformed rock segments called microlithons.

Strain Amount of deformation.

Strain shadows Roughly triangular areas of quartz or another mineral on either side of a rigid grain, such as a garnet porphyroblast. It results from rock extension in the direction of the strain shadows.

Stratigraphy The layered sequence of a particular region. The sequence may be chronological, like a sequence of undeformed sedimentary or volcanic rock layers (stratigraphic sequence). The sequence may also be tectonic (tectonic stratigraphy or tectonostratigraphy), where there may be repeated and/or overturned layers caused by thrust faults and overturned folds, but nonetheless arranged in the same or similar ways over a broad region (e.g., the Lower, Middle, Upper, and Uppermost Allochthons of western Scandinavia).

Stress Applied force.

Strike A horizontal line on an inclined surface, such as a cleavage surface. The strike line is perpendicular to the dip direction.

Subgrains Deformed crystals commonly develop small domains that are obviously derived from one crystal, but with the crystal lattice rotated or otherwise distorted by a small amount (easily visible in quartz, for example, in thin section with a polarizing microscope).

Subhedral Between euhedral and anhedral, a crystal with some facets.

Supercritical fluid A fluid at a temperature and/or pressure above its critical point. Above the critical point there is no distinct transition between high-density and low-density fluid as the pressure and/or temperature changes.

Supersaturated A solid, liquid, or gas that contains a sufficient concentration of something for precipitation to occur, but precipitation does not take place because there are no suitable nuclei available where precipitation can start, or because the processes that are necessary for precipitation are too slow.

Symplectite An intergrowth of minerals where the mineral shapes are worm-like, lath-like, or in some other semi-regular arrangement. These typically form from the reaction of one mineral to others, such as omphacite to augite plus plagioclase in retrograded eclogites.

Syncline In less-deformed regions this refers to a fold arched (convex) downwards. In more deformed regions, with overturned folds, the word is restricted to folds that have progressively older rocks on the convex side.

Synform A convex-downward fold in which either the relative age of the different layers is unknown, or in which the layer age sequence varies because of folding or thrust faulting.

T Temperature.

Tail A continuous line of recrystallized mineral grains extending to either side of a porphyroclast, or a similar-looking thin rock extension from boudin edges.

Tectonic Referring to the large-scale movement of material within the Earth, to produce folds, faults, boudins, moving plates, and so on.

Tectonostratigraphy A more or less regularly layered sequence, like a sedimentary stratigraphy, but assembled at least partly by tectonic processes such as the stacking of rock sequences on top of one another along thrust faults.

Terminated An elongate crystal with facets on at least one end.

Thermodynamics The science and mathematics of heat, heat flow, and work. It is used for everything from steam engines to predicting mineral assemblages and compositions in metamorphic rocks. Very useful.

Thin section A thin slice of rock, usually about 30 μm thick (30 microns), glued to a glass slide. The exposed side is usually polished or covered with a thin glass coverslip. Most minerals are transparent at this thickness, which allows the mineralogy and textures to be observed in great detail with optical and microbeam techniques.

Thrust fault An inclined fault where the upper block moved upwards relative to the lower block. The term 'reverse fault' is used by some for high-angle (steep) thrust faults.

Transposed bedding In real bedding, the original contacts between distinct sedimentary or volcanic layers are intact and continuous. During deformation, cleavage surfaces tend to offset the bedding along closely-spaced shear planes. From a distance, the bedding surface may look intact, but in fact it has become discontinuous, or transposed.

Trench The elongate regions of very deep water (up to 11 km) where oceanic plates bend downward and descend into the mantle. They are the most oceanward parts of the subduction zone-volcanic arc system.

Triassic Geologic period in the Mesozoic Era, between about 252 and 201 million years ago

Turbidite A sedimentary layer that has been deposited by a turbidity current. Those form from submarine landslides of unconsolidated material. As the material flows downhill, it mixes with water to become a turbulent slurry (a turbidity current).

As the current slows on the flat topography in deep water, it deposits its detrital load, coarse particles first, followed by smaller particles, producing a graded bed. This deposit is a turbidite.

Unconformity An erosion surface, buried under sedimentary or volcanic rock. It remains an unconformity if it is metamorphosed.

Unroofing See exhumation.

Vein A fracture that has been filled with minerals precipitated by flowing fluid.

Vermicular Worm-like in shape, such as crystals in many symplectites.

Vesicles Gas bubbles in a volcanic or shallow plutonic rock. If filled with minerals, such as calcite or zeolites, they are instead called amygdules (or amygdales).

Volatiles Components in a rock, like H_2O and CO_2, that can be released as gasses at low-pressure.

Volcanic Related to magma that solidifies on the Earth's surface, including lava flows, ash and other loose pyroclastic fragments, lava dome collapse deposits, and so on.

Volcaniclastic A sedimentary deposit made mostly of transported volcanic material.

Xenocryst A mineral grain engulfed in an unrelated igneous rock. For example, a crystal of quartz broken from the host rock of a basaltic dike.

Xenolith A block of rock engulfed in an unrelated igneous rock, for example a block of marble in basalt, or a block of Archean gneiss in much younger granite.

Zoning In a mineral grain it refers to chemical and possibly related color variations, or variations in inclusion abundance, from one part to another (e.g., watermelon tourmaline or garnets with Mn-rich cores and Mg-rich rims). In rocks it refers to orderly variations in mineralogy or mineral proportions, usually radially from some center (e.g., mineralogy zoning in contact metamorphic aureoles).

GLOSSARY OF ROCK TYPES

Rock types mentioned in this book. Each is noted as being metamorphic, sedimentary, plutonic (igneous), or volcanic (igneous). In other contexts the definitions may be somewhat different.

Amphibolite Metamorphic, derived from basaltic or andesitic rock, made mostly of hornblende and plagioclase.

Andesite Volcanic, having intermediate silica, K, Na, Mg, Fe, and Ca contents, typical of subduction zone volcanoes. Equivalent to diorite.

Anorthosite Plutonic, composed of 90% or more plagioclase (70-90% is gabbroic anorthosite). These are generally thought to be cumulate rocks.

Aplite Plutonic, generally dikes of granitic composition having a fine-grained, granular texture.

Arkose Sedimentary, feldspar-rich sandstone. Because feldspars are unstable and weather relatively quickly, arkoses are generally found close to the sediment source areas.

Basalt Volcanic, low in silica, K, and Na, and rich in Mg, Fe, and Ca. The most common volcanic rock type. Equivalent to gabbro.

Blueschist Metamorphic, derived from basalt or andesite, forming at high-pressure and relatively low-temperature in subduction zones. It is dominated by the blue mineral glaucophane.

Breccia Sedimentary, igneous, or metamorphic, a rock made of or containing abundant angular fragments of any precursor rock.

Calc-silicate rock Metamorphic, dominated by calcium-rich silicate minerals such as diopside and grossular. These can be derived from sediments such as impure carbonate rocks, marls, or mixtures of volcaniclastics and carbonate. They can also be formed in veins, and in reaction zones between carbonate rocks and plutons.

Carbonatite Volcanic and plutonic, made mostly of carbonate minerals, very rare.

Charnockite Plutonic, an orthopyroxene-bearing granite.

Chert Sedimentary, made of microcrystalline quartz.

Clinopyroxenite Metamorphic or plutonic, made mostly of clinopyroxene such as augite. Igneous pyroxenites are cumulates. Metamorphic pyroxenites may be metamorphosed cumulates, or particularly pyroxene-rich calc-silicate rocks.

Conglomerate Sedimentary, made mostly of grains more than 2 mm in diameter (pebbles, cobbles, gravel).

Coticule Metamorphic, an unusual quartzite with a high proportion of Fe- or Mn-rich garnet. These pink rocks are interpreted to be metamorphosed Fe- or Mn-rich chert.

Cumulate rock Plutonic, formed from the segregation of crystals from a magma, such as by the crystals settling to the magma chamber floor to form layers.

Dacite Volcanic, relatively high silica, K, and Na, and low Mg, Fe, and Ca contents, in between rhyolite and andesite. Equivalent to granodiorite or tonalite.

Diorite Plutonic, made mostly of plagioclase and mafic minerals with little quartz or K-feldspar. Equivalent to andesite.

Dolostone Sedimentary, a carbonate rock composed mostly of dolomite.

Dunite Plutonic or metamorphic, made of more than 90%f olivine. These are igneous cumulates, or metamorphosed cumulates.

Eclogite Metamorphic, made mostly of red garnet and green omphacite. It is derived from basaltic or andesitic rocks and formed at high-pressure, moderate-temperature conditions in subduction zones.

Gabbro Plutonic, dominated by plagioclase and pyroxene. Equivalent to basalt.

Garbenshiefer Metamorphic, containing radiating sprays of hornblende, like sheaves of wheat. Muscovite, biotite, carbonate, chlorite, quartz, and garnet are also commonly present. Derived from a sedimentary mixture of mostly shale and basalt volcaniclastic components.

Gneiss Metamorphic, poorly-foliated, mica-poor, feldspar-rich, and usually relatively coarse-grained. Gneisses are generally derived from intermediate or felsic plutonic or volcanic rocks, or others such as shales that have been metamorphosed at high enough grade to dehydrate most of the micas.

Granite Plutonic, made of approximately equal proportions of K-feldspar, plagioclase, and quartz. Equivalent to rhyolite.

Granodiorite Plutonic, made of approximately equal proportions of plagioclase and quartz, with less K-feldspar than granite but more than tonalite. Equivalent to dacite.

Granofels Metamorphic, medium- to coarse-grained with a granular texture, unfoliated and unlineated. Similar to a granulite, but the term is more commonly used in the context of contact metamorphic rocks.

Granulite Metamorphic, medium- to coarse-grained with a granular texture, unfoliated and unlineated. Similar to granofels. This textural term should not be confused with the pyroxene granulite metamorphic facies.

Greenschist Metamorphic, low-grade and green, commonly derived from basalt or andesite. Generally foliated with some combination of white mica, chlorite, actinolite, and epidote.

Harzbergite Plutonic or metamorphic, made mostly of olivine and enstatite. Plutonic varieties are cumulates. Metamorphic varieties may be restite from partial melting of mantle peridotite, or metamorphosed cumulates.

Hornblendite Plutonic or metamorphic, made mostly of hornblende. Plutonic hornblendites are cumulates. Metamorphic varieties may be metamorphosed igneous hornblendites, metamorphosed cumulates made of a chemically equivalent mixture of anhydrous minerals, or metamorphosed basaltic rocks of the right composition and metamorphic grade.

Hornfels Metamorphic, generally fine-grained, unfoliated, unlineated, and flinty, usually associated with contact metamorphic zones.

Kimberlite Plutonic or volcanic, formed form magmas derived from deep in the upper mantle. Generally K- and Mg-rich, diamond-bearing in some cases. Very rare.

Marble Metamorphic, derived from limestone or dolostone.

Marl Sedimentary, composed principally of a mixture of clay and carbonate minerals.

Monzonite Plutonic, made of approximately equal proportions of K-feldspar and plagioclase, with no or just small quantities of quartz (much less than granite).

Mylonite Metamorphic, fault rock formed by grain size reduction during ductile faulting.

Nepheline syenite Plutonic, a felsic rock much like granite but lower in silica and higher in Na and/or K, with nepheline present instead of quartz.

Norite Plutonic, composed mostly of plagioclase and orthopyroxene. These are generally cumulates.

Pegmatite Plutonic, generally dike- or sill-like bodies formed of very coarse-grained feldspar, quartz, and other minerals that crystallized from H_2O-rich silicate liquid. Some pegmatites may have been precipitated by hot, high-pressure aqueous solutions of the same minerals. Pegmatites may be offshoots from cooling felsic plutons, or may form by partial melting and melt separation during medium- to high-grade metamorphism.

Pelitic rocks Metamorphic, formed from shale.

Peridotite Plutonic or metamorphic, made mostly of olivine and pyroxene. Plutonic varieties are cumulates. Metamorphic varieties may be metamorphosed cumulates, or they may always have been peridotites as part of the upper mantle.

Phyllite Metamorphic, well-foliated and fine-grained, with sheet silicates generally too small to identify but coarse enough to give rock cleavage surfaces a satin-like reflectivity. Most phyllites were derived from shales, but other protoliths are possible.

Pseudotachylite Metamorphic, produced by rapid fault movement. These are dark, flinty rocks that are usually found in fractures of limited extent in fault zones. They are made of very finely crushed rock and possibly devitrified glass, that may enclose larger fragments of the host rock.

Pyroclastic Volcanic, made of fragmented, solidified magma (scoria, volcanic ash, tephra, pyroclastic flow deposits, etc.).

Pyroxene granulite Metamorphic, a granular rock made of orthopyroxene, clinopyroxene, plagioclase, and smaller amounts of other minerals. Usually derived from rocks of basaltic or andesitic composition. Indicative of pyroxene granulite facies metamorphic grade.

Pyroxenite Igneous or metamorphic, made mostly of pyroxene (ortho- and/or clino-pyroxene). The igneous varieties are cumulates. The metamorphic varieties may be metamorphosed cumulates, otherwise most are particularly pyroxene-rich calc-silicate rocks.

Quartzite A metamorphosed sedimentary rock composed mostly of quartz.

Rapakivi granite Plutonic, a granite (obviously) containing feldspar phenocrysts that have K-feldspar cores and albite rims.

Rhyolite Volcanic, rich in silica, K, and Na, poor in Mg, Fe, and Ca. Equivalent to granite.

Schist Metamorphic, relatively coarse-grained and well-foliated with abundant sheet silicates, such as muscovite, biotite, chlorite, or talc. Most schists have shale protoliths.

Serpentinite Metamorphic, rich in serpentine minerals, usually derived from olivine-rich ultramafic rocks. They commonly also contain talc, carbonates, chlorite, and magnetite.

Shale a fine-grained sedimentary rock composed primarily of clay. Quartz and carbonates are other common constituents.

Skarn Metamorphic, made mostly of calc-silicate minerals like diopside and grossular garnet. Analogous to calc-silicate rocks, but more typically refers to masses at contact zones between igneous plutons and carbonate rocks.

Slate Metamorphic, fine-grained, easily cleaved, derived from shale. Generally rich in chlorite, white mica, and quartz, but grains are too small to be visible in the field.

Tonalite Plutonic, made mostly of quartz and plagioclase with almost no K-feldspar. Equivalent to dacite.

Ultramafic rock Igneous or metamorphic, a rock made of more than 90% mafic minerals. Most igneous varieties are cumulates. Metamorphic varieties may be metamorphosed cumulates or mantle rocks. Some basaltic rock compositions can effectively become ultramafic rocks at some metamorphic grades. Many calc-silicate rocks are technically ultramafic, but are usually referred to as calc-silicate rocks as a reminder of their calcareous heritage.

Websterite Plutonic or metamorphic, made mostly of clinopyroxene and orthopyroxene. Plutonic varieties are cumulates. Metamorphic varieties are probably metamorphosed cumulates, or possibly restite from partially melted pyroxene granulite.

GLOSSARY OF MINERALS

Amphiboles, micas, and clays, among other minerals, come in different varieties. In their crystal lattices there are particular sites that, in some varieties, atoms occupy the sites, whereas in others the sites are empty (vacancies, □). Edenite, for example, is an amphibole variety with the formula, $NaCa_2(Mg,Fe^{2+})_5(Si_7,Al)O_{22}(OH)_2$. Na fully occupies a particular large site in the crystal lattice. In contrast, actinolite, another amphibole, has the formula $\square Ca_2(Mg,Fe^{2+})_5Si_8O_{22}(OH)_2$. In tremolite, the large site is empty, or vacant. Vacancies were omitted in the rest of the book because it was the minerals, and their part in chemical reactions, that were important. Here, the vacancies are included to better show the systematic structural and chemical relationships between members of mineral groups.

Actinolite □$Ca_2(Mg,Fe^{2+})_5Si_8O_{22}(OH)_2$ Amphibole group, monoclinic, generally long, green crystals.

Albite $NaAlSi_3O_8$ A feldspar in the plagioclase subgroup, generally white, blocky crystals, found in felsic igneous rocks and a wide variety of metamorphic rocks.

Aluminosilicate Al_2SiO_5 A mineral group that includes kyanite, sillimanite, and andalusite, also mullite in some contact metamorphic rocks.

Amphibole A group of silicate minerals with double chains of silica tetrahedra. All amphiboles have two cleavages that intersect at about 56° and 124° angles, and collectively have a broad range of chemical composition. Examples are actinolite and hornblende.

Analcite $NaAlSi_2O_6 \bullet (H_2O)$ A zeolite-group mineral, white, equant crystals in very low-grade, low-pressure rocks and hydrothermal veins.

Andalusite Al_2SiO_5 One of the aluminosilicate minerals, usually white, pink, or gray orthorhombic crystals with two cleavages at about 90°.

Andradite $Ca_3Fe^{3+}{}_2Si_3O_{12}$ Garnet group, characteristic of Ca-rich rocks, equant crystals, usually colored orange, red, brown, or black.

Ankerite $CaFe^{2+}(CO_3)_2$ Rhombohedral carbonate, crystals generally colorless or tan, rusty-weathering.

Anorthite $CaAl_2Si_2O_8$ A feldspar in the plagioclase group, generally white, blocky crystals, usually found in Ca-rich rocks.

Anthophyllite □$(Fe^{2+},Mg)_2(Mg,Fe^{2+})_5Si_8O_{22}(OH)_2$ Amphibole group, orthorhombic, generally long, colorless, green, gray, or brown crystals.

Antigorite $(Mg,Fe^{2+})_3Si_2O_5(OH)_4$ Serpentine group, a soft sheet silicate, usually colorless, gray, or green.

Apatite $Ca_5(PO_4)_3(OH,F,Cl)$ The most common phosphate mineral in metamorphic and igneous rocks, usually minute, hexagonal, colorless, yellow, green, or blue crystals.

Aragonite $CaCO_3$ Orthorhombic carbonate mineral, found in low-temperature, high-pressure rocks like blueschists, usually colorless or gray.

Augite $(Ca,Fe^{2+},Mg,Na)(Mg,Fe^{2+},Al,Fe^{3+},Ti)(Si,Al)_2O_6$ A clinopyroxene solid solution typical of igneous rocks. Augite also occurs in some granulite facies mafic and intermediate rocks.

Biotite $K(Mg,Fe^{2+},Al,Ti,Fe^{3+})_3(Si,Al)_4O_{10}(OH)_2$ Mica group, usually black with a single perfect cleavage.

Calcite $CaCO_3$ The most common rhombohedral carbonate, crystals usually colorless.

Carbonate XCO_3 Minerals having $CO_3{}^{2-}$ groups as an important part. These minerals fizz in acid, though some do so very slowly. Also refers to rocks that have carbonate minerals as the major part.

Chalcopyrite $CuFeS_2$ A brass-yellow, metallic sulfide mineral.

Chlorite $(Mg,Fe^{2+},Al)_3(Si,Al)_4O_{10}(OH)_2 \bullet (Mg,Fe^{2+},Al)_3(OH)_6$ Soft, platy mineral with one perfect cleavage, usually green.

Chloritoid $(Fe^{2+},Mg,Mn^{2+})_2Al_4Si_2O_{10}(OH)_4$ Hard, blocky to platy grains, usually dark-green to black.

Clay A class of sheet silicates having structures like the micas or serpentine-group minerals, but with grain sizes generally less than 10 μm across. As a group, clays have a wide range of chemical compositions that can include molecular H_2O. They are produced during chemical weathering, hydrothermal alteration, and other near-surface processes.

Clinopyroxene Monoclinic pyroxene, here used to refer to diopside, hedenbergite, or augite, but in other circumstances can refer to omphacite and others.

Clinozoisite $Ca_2AlAl_2(SiO_4)(Si_2O_7)O(OH)$ Epidote group, monoclinic, usually elongate colorless, gray, or yellow-green crystals.

Cordierite $(Mg,Fe^{2+})_2Al_4Si_5O_{18}$ Ring silicate found in medium- to high-grade, moderate- to low-pressure peraluminous rocks, crystals usually irregular or blocky, usually gray, blue, or violet.

Chrysotile $(Mg,Fe^{2+})_3Si_2O_5(OH)_4$ Serpentine group, but a fibrous variety, rather than platy. Usually white, gray-green, or green.

Cummingtonite $\square(Fe^{2+},Mg)_2(Mg,Fe^{2+})_5Si_8O_{22}(OH)_2$ Amphibole group, monoclinic, generally long, gray, gray-brown, or gray-green crystals.

Diamond C Cubic form of carbon, produced in rocks that have experienced very high pressures.

Diopside $Ca(Mg,Fe^{2+})Si_2O_6$ Clinopyroxene group, usually found in Ca-rich rocks, forms blocky crystals that can be colorless, light-green, or dark-green.

Dolomite $CaMg(CO_3)_2$ Rhombohedral carbonate, crystals usually colorless or tan.

Enstatite $(Mg,Fe^{2+})(Mg,Fe^{2+})Si_2O_6$ Orthopyroxene group, usually gray-green to gray-brown blocky crystals, rusty-weathering.

Epidote $Ca_2(Fe^{3+},Al)Al_2(SiO_4)(Si_2O_7)O(OH)$ Epidote group, monoclinic, blocky to elongate crystals, usually yellow-green to dark-green.

Feldspar Most abundant mineral, as a group, in Earth's crust. All are K-, Na-, or Ca-bearing framework aluminum silicates, with related monoclinic or triclinic structures. They have two cleavages that intersect at approximately 90°, and occur in a variety of colors, usually white, gray, or pink.

Feldspathoids A group of Na, K, and Ca silicate minerals that have a passing resemblance to feldspars. Generally colorless or gray, with relatively low refractive indices. Includes minerals like nepheline, kalsilite, and scapolite.

Forsterite $(Mg,Fe^{2+})_2SiO_4$ Olivine group, blocky crystals, uncommon in metamorphic rocks except marble and ultramafic rocks at medium- to high-grade, and silica-poor, high-temperature contact metamorphic rocks. Usually colorless, yellow-green, or green. Rusty-weathering. Common in igneous rocks.

Garnet $(Fe^{2+},Mg,Ca,Mn^{2+})_3(Al,Fe^{3+})_2Si_3O_{12}$ A group of hard, isometric minerals, lacking cleavage, with a wide solid solution range. The common red garnet us usually an Fe-rich solid solution $(Fe^{2+},Mg,Ca,Mn^{2+})_3Al_2Si_3O_{12}$. In marbles and calc-silicate rocks, garnets tend to be $Ca_3(Al,Fe^{3+})_2Si_3O_{12}$ solid solutions.

Gedrite $(\square,Na)(Fe^{2+},Mg)_2(Mg,Fe^{2+},Al)_5(Si,Al)_8O_{22}(OH)_2$ Amphibole group, orthorhombic, generally long, gray, brown, greenish, or black crystals.

Glaucophane $\square Na_2((Mg,Fe^{2+})_3(Al,Fe^{3+})_2)Si_8O_{22}(OH)_2$ Amphibole group, found in blueschist facies high-pressure, low-temperature rocks, usually blue or blue-gray.

Graphite C Hexagonal form of carbon, metallic gray, very soft with one perfect cleavage.

Grossular $Ca_3Al_2Si_3O_{12}$ Garnet group, common in Ca-rich rocks like marble, usually equant crystals, colorless, green, yellow, gray, brown, or pink.

Hedenbergite $Ca(Fe^{2+},Mg)Si_2O_6$ Clinopyroxene group, usually found in Ca-rich rocks, blocky crystals that are dark-green to black.

Hematite Fe_2O_3 Rhombohedral ferric iron oxide, metallic gray, hard mineral if crystals are large, but if they are micron-scale they can color rocks and minerals brick red or pink. The pink color common in microcline is caused by unmixing of Fe^{3+} from solid solution to form minute hematite plates.

Hornblende $(\square,Na,K)(Ca,Na,Fe^{2+},Mg)_2(Mg,Fe^{2+},Fe^{3+},Al,Ti)_5(Si,Al)_8O_{22}(OH)_2$ Amphibole group, monoclinic, the field name for common black, monoclinic amphiboles. If the chemical composition becomes known, the field name may change based on standard naming conventions (Leake et al., 1997).

Illite $(\square,K,Na,H_2O)(Al,\square,Mg,Fe)_3(Si,Al)_4O_{10}(OH)_2$ A clay mineral having mica-like structure, usually white, gray, or green, but crystals are generally only microns across.

Ilmenite $FeTiO_3$ Rhombohedral iron-titanium oxide, hard, metallic gray.

Jadeite $Na(Al,Fe^{3+})Si_2O_6$ Clinopyroxene group, characteristic of high-pressure, low-temperature metamorphism of Na-rich rocks, usually white or green, fibrous to blocky crystals.

Kalsilite $KAlSiO_4$ One of the 'feldspathoids', rare, found only in silica-poor, high-temperature, usually igneous rocks.

Kaolinite $(\square Al_2)Si_2O_5(OH)_4$ A clay mineral with a structure like serpentine, usually white.

K-feldspar $KAlSi_3O_8$ A general term referring to the K-rich feldspars that include microcline, orthoclase, and sanidine, which differ somewhat in structure and optical properties. The term can also be used for K-Na feldspar solid solutions, and perthite, particularly in the field.

K-mica Any of the K-rich micas, muscovite, biotite, phlogopite, or phengite.

Kyanite Al_2SiO_5 One of the aluminosilicate minerals, triclinic, usually blade-shaped, colorless, gray, or blue. Has one perfect cleavage and two fair cleavages that intersect at large angles (not quite 90°).

Labradorite $(Ca,Na)(Si,Al)_4O_8$ A feldspar in the plagioclase group with Ca:Na in the proportion range of 50:50 to 70:30. Crystals usually blocky, colorless to gray.

Lawsonite $CaAl_2Si_2O_7(OH)_2\bullet(H_2O)$ A hydrous silicate found in very low-temperature, moderate- to high-pressure rocks, usually small, blocky crystals, colorless, gray, or blue.

Limonite A mineraloid having no well-defined crystal structure. Its composition varies but it is approximately between $Fe^{3+}(OH)_3$ and $Fe^{3+}O(OH)$, with bound molecular H_2O. It is the yellow to brown rusty staining seen widely on outcrops.

Magnesite $MgCO_3$ Rhombohedral carbonate typically found in low- to medium-grade ultramafic rocks. Usually colorless.

Magnetite Fe_3O_4 Magnetic iron oxide, forms equant, metallic, dark-gray crystals.

Mica A group of sheet silicates having a 3-layer structure, K, Na, or Ca as interlayer ions, and one perfect cleavage. Examples are muscovite and biotite.

Microcline $KAlSi_3O_8$ Feldspar group, triclinic, low-temperature form of K-feldspar. Usually blocky crystals that can be white, gray, or pink.

Monazite $(LREE)PO_4$ A common phosphate mineral in peraluminous rocks like pelitic schist (LREE stands for Light Rare Earth Elements, meaning low atomic number lanthanides, which have larger ionic radii than the rest).

Mullite A high-temperature, low-pressure, defect form of sillimanite (Al_2SiO_5) with the formula $Al_2(Si_{1-x},Al_x)O_{5-(x/2)}$, $x=0.15\text{-}0.6$.

Muscovite $K(\square Al_2)Si_3AlO_{10}(OH)_2$ Mica group, the common 'white mica', generally colorless or gray with one perfect cleavage.

Nepheline $(Na,K)AlSiO_4$ One of the 'feldspathoids', found in metamorphosed nepheline syenite. A possible mineral in high-temperature, silica-poor carbonate rocks. Crystals are generally blocky, colorless or gray.

Olivine $(Mg,Fe^{2+})_2SiO_4$ Olivine group, blocky crystals found in ultramafic rocks, basalts, and other silica-poor, Mg- or Fe-bearing rocks like marbles. Usually colorless, yellow-green, or green.

Omphacite $(Ca,Na)(Mg,Fe^{2+},Al,Fe^{3+})Si_2O_6$ Clinopyroxene group, intermediate between diopside and jadeite. It is the characteristic green pyroxene found in eclogites.

Orthoamphibole Orthorhombic variety of amphibole group minerals that include anthophyllite and gedrite. They have the same double silicate chain structure, two cleavages, and elongate crystal forms as the monoclinic amphiboles.

Orthoclase $(K,Na)AlSi_3O_8$ Feldspar group, monoclinic, blocky crystals that are usually colorless or gray. Intermediate temperature form of K-feldspar.

Orthopyroxene $(Fe^{2+},Mg)(Mg,Fe^{2+})Si_2O_6$ Pyroxene group, orthorhombic, but having the same single silicate chain structure as the monoclinic pyroxenes. Orthopyroxenes generally weather faster than clinopyroxenes, and so commonly have a rusty stain on them on outcrop surfaces.

Osumilite $(K,Na)(Mg,Fe^{2+})_2(Al,Fe^{3+})_3(Si,Al)_{12}O_{30}$ A double-ring silicate found in high-temperature, low-pressure, aluminous metamorphic rocks. Crystals typically blocky, usually brown, green, dark-blue, or black.

Paragonite $Na(\square Al_2)Si_3AlO_{10}(OH)_2$ Mica group, the Na-analog of muscovite, characteristically found in low-grade rocks, replaced by albite at higher grade. Generally indistinguishable from muscovite in the field. Lore has it that, for mica powder scratched off a cleavage surface, paragonite floats on beer, but muscovite sinks.

Pargasite $NaCa_2(Mg,Fe^{2+})_4Al(Si_6Al_2)O_{22}(OH)_2$ Monoclinic amphibole, green to black. A widespread solid solution component in field-identified 'hornblende'. Characteristic of high-pressure and high-temperature metamorphic rocks.

Perthite $(K,Na)AlSi_3O_8$ A special texture that forms where a high-temperature, homogeneous K-Na feldspar solid solution unmixes at low-temperature to form alternating lamellae of K-rich microcline or orthoclase and Na-rich albite. Where coarse, the lamellae are easily visible in hand sample.

Phengite $K(\square,Mg,Fe^{2+},Al)_2)Si_4O_{10}(OH)_2$ Mica group, a 'white mica' like muscovite, but characteristic of high-pressure rocks such as blueschist or eclogite.

Phlogopite $KMg_3Si_3AlO_{10}(OH)_2$ Mica group, characteristic of ultramafic rocks and marble. It has the single perfect cleavage of all micas, but is pale-colored (colorless, gray, brown, red-brown) because of its low iron content.

Plagioclase Feldspar group, metamorphic varieties are all triclinic, encompassing the composition range $NaAlSi_3O_8$ (albite) to $CaAl_2Si_2O_8$ (anorthite). At high-temperature they form a complete solid solution series, but at low temperatures they may not. In some rocks two plagioclase feldspar compositions can coexist in the same assemblage. Very little K can enter in solid solution with Ca-rich plagioclase at any temperature, but at high-temperature the Na-rich plagioclases can form a complete Na-K feldspar solid solution series.

Prehnite $Ca_2Al_2Si_3O_{10}(OH)_2$ Characteristic of low-temperature veins and metamorphic rocks. Crystals are typically blocky and colorless to light-green.

Pumpellyite $Ca_2(Al,Fe^{2+},Mg)Al_2(SiO_4)(Si_2O_7)(OH,O)_2 \bullet H_2O$ A platy mineral characteristic of low-temperature veins and metamorphic rocks, usually colorless.

Pyrite FeS_2 A common light-yellow, metallic sulfide.

Pyrope $Mg_3Al_2Si_3O_{12}$ Garnet group, found in high-pressure, usually ultramafic rocks, equant crystals, typically pink or red.

Pyrophyllite $\square(\square Al_2)Si_4O_{10}(OH)_2$ A very soft sheet silicate mineral like talc and the micas, found in hydrothermal veins and low- to moderate-temperature aluminous metamorphic rocks.

Pyroxene A group of minerals with related monoclinic or orthorhombic structures, all having single silicate tetrahedral chains and two cleavages that intersect at angles close to 90°.

Pyrrhotite $Fe_{1-x}S$ ($x=0$-0.17) A common bronze-yellow, metallic, and magnetic sulfide mineral in medium- to high-grade metamorphic rocks. Tends to weather much faster than pyrite or chalcopyrite.

Quartz SiO_2 The common silica mineral, found in a wide range of rocks. Crystals are usually irregular (except in open veins), glassy clear to gray, hard, no cleavage.

Riebeckite $\square Na_2(Fe^{2+}{}_3Fe^{3+}{}_2)Si_8O_{22}(OH)_2$ Amphibole group, usually found in oxidized, Na-rich rocks. Crystals are usually elongate, blue to black.

Rutile TiO_2 The common titanium oxide, typically found in high-pressure rocks, and some intermediate pressure rocks, especially those that are Mg-rich. Crystals are highly reflective, elongate to blocky, typically black or brown if large, orange or red if very small.

Sanidine $(K,Na)AlSi_3O_8$ Monoclinic, high-temperature form of K-rich feldspars. Found in quickly-cooled volcanic or high-temperature contact metamorphic rocks. Inverts by atomic rearrangement to orthoclase or microcline if cooled too slowly.

Scapolite $Na_4Al_3Si_9O_{24}Cl$ to $Ca_4Al_6Si_6O_{24}CO_3$ The Na-rich varieties can be found in a wide range of Cl^-- and CO_3^{2-}-bearing veins and metamorphic rocks, whereas the Ca-rich varieties are found mostly in high-temperature, Ca-rich rocks.

Sericite Fine-grained white mica, such as that in phyllite, or as a product of feldspar alteration.

Serpentine $(Mg,Fe^{2+})_3Si_2O_5(OH)_4$ A mineral group, all of which are two-layer sheet silicates with a single perfect cleavage. The group includes chrysotile (fibrous), antigorite, and lizardite (both platy). These are found in low- to medium-grade metamorphosed ultramafic rocks, and low-grade, retrograded, or hydrothermally altered olivine-bearing rocks.

Sillimanite Al_2SiO_5 One of the aluminosilicate minerals, orthorhombic, usually needle-shaped, white or gray crystals with one cleavage parallel to the length, and common cross-fractures. Usually occurs as very small fibers or needles.

Spinel $(Mg,Fe^{2+})Al_2O_4$ A hard oxide mineral found in silica-poor, high-temperature metamorphic rocks. Typical crystals are shaped like octahedra, usually colorless or green.

Staurolite $(Fe^{2+},Mg,Zn,\square)_2(Al,Fe^{3+},Mg,Ti)_9O_6[(Si,Al)O_4]_4(O,OH)$ A hard silicate mineral that usually forms blocky to elongate, brown crystals, common in medium-grade pelitic schists but also found in some low-Ca mafic rocks.

Stilpnomelane $K(Fe^{2+},Mg,Fe^{3+})_8(Si,Al)_{12}(O,OH)_{27} \bullet 2(H_2O)$ A black sheet silicate, not of the mica- or serpentine-groups, characteristic of low-temperature and Fe-rich rocks.

Sulfur S A soft, yellow mineral found in reduced sedimentary rocks and in deposits around hydrothermal vents.

Talc □$(Mg,Fe^{2+})_3Si_4O_{10}(OH)_2$ A very soft sheet silicate having mica-like structure, usually found in low- and medium-grade ultramafic rocks. It has one perfect cleavage, and is usually white, tan, or light-green.

Titanite $CaTiSiO_5$ A common Ti-rich mineral in low- and medium-grade metamorphic rocks, highly reflective, characteristically double wedge-shaped crystals, usually yellow, brown, or black.

Tourmaline $Na(Fe^{2+},Mg)_3Al_6(Si_6O_{18})(BO_3)_3(OH)_3(F,OH)$, the common black variety. Typically forms stubby to elongate prisms, most common in veins and in low- to medium-grade pelitic phyllites and schists, but can occur in other metamorphic rocks if they have been metasomatically enriched in boron.

Tremolite □$Ca_2Mg_5Si_8O_{22}(OH)_2$ Amphibole group, monoclinic, generally long, colorless or gray crystals, characteristically found in low- to medium-grade marbles and calc-silicate rocks.

Tridymite SiO_2 A high-temperature form of silica, preserved in some quickly-cooled, high-temperature contact metamorphic rocks.

White mica Generally referring to the nominally colorless micas like paragonite, phengite, and muscovite.

Wollastonite $CaSiO_3$ A single-chain silicate, not of the pyroxene-type, usually found in calc-silicate rocks and marbles that have been metamorphosed to high-temperature and relatively low-pressure.

Zeolite A large class of framework aluminum silicates in which there are large spaces in the silicate frame, into which charge-balancing ions like K, Na, and Ca can enter, along with molecular water and other small molecules. Common in a wide range of low-temperature, low-pressure veins and metamorphic rocks.

Zircon $ZrSiO_4$ The most common zirconium mineral. Generally too small to be seen in the field, but where large it forms brownish, highly reflective, tetragonal crystals.

Zoisite $Ca_2AlAl_2(SiO_4)(Si_2O_7)O(OH)$ Epidote group, orthorhombic, usually elongate colorless, gray, or yellow-green crystals, typically found in calc-silicate rocks, some marbles, Al-rich eclogites, and as an alteration product of Ca-rich plagioclase.

Index

Printed and bound by CPI Group (UK) Ltd, Croydon, CR0 4YY
18/10/2024
01776249-0002